Safety Leadership

Safety Leadership
A Different, Doable and Directed Approach to Operational Improvements

Robert J. de Boer

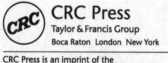

CRC Press
Taylor & Francis Group
Boca Raton London New York

CRC Press is an imprint of the
Taylor & Francis Group, an **informa** business

First edition published 2021
by CRC Press
6000 Broken Sound Parkway NW, Suite 300, Boca Raton, FL 33487-2742

and by CRC Press
2 Park Square, Milton Park, Abingdon, Oxon, OX14 4RN

First edition published by CRC Press 2021

CRC Press is an imprint of Taylor & Francis Group, LLC

ISBN: 9780367698089 (hbk)
ISBN: 9780367652753 (pbk)
ISBN: 9781003143338 (ebk)

Typeset in Palatino
by codeMantra

For my parents.

Contents

Foreword

'Nobody reads anything anymore', my old professor (David Woods) told me grumpily one day. This may well be true, even though the statement would seem a bit sweeping and generalising. Yet, if it is, then it is hugely disheartening for somebody who writes a lot of books. Of course, we are all time-poor, and the competition for our attention by all kinds of other media – not books – is probably stronger than it's ever been.

With Robert J. de Boer's new book, however, you no longer have an excuse. If ever there was a low-threshold book to get you into the world of Safety Differently, and Safety II, then this is it. There are a bunch of things that give this book its low threshold, so that you can easily step over it and start learning what these things are about; for you to start talking the talk; for you to begin to have a better notion of what the safety types in your own organisation (hopefully) are talking about.

The first feature – virtue, really – that makes this book one that you cannot afford not to read, is that it's short. Writing a short book is harder than writing a long one, a thick one, a tome. Reading a short book is easier. Of course, it will leave you, hopefully, longing for more. But that's no problem. There's always more to read, more to learn. There are other books on Safety Differently and Safety II that Robert points you to, and that you can get into once you've got this one digested.

The second feature is that it is comprehensive. This little book really hits all the important talking points relevant to safety today. Psychological safety, welcoming bad news, avoiding retribution, sensemaking and local rationality, Work-as-Imagined versus Work-as-Done, complexity, micro-experiments, drifting into failure, safety margins, and how to maximise learning after something has gone wrong – it's all there. Robert has taken care to be inclusive, yet not laboriously so. This little book hits the sweet spot between what you really need to know and what you have time for.

The third feature is, quite nifty, a glossary. If you're not (so) familiar with safety, or with the Safety Differently or Safety II approach to it, then having a good overview of the verbiage, the phrases used, the contemporary 'patois' of safety, is key. With this in hand, you can more swiftly ease into the conversations that matter. And you can avoid asking stupid questions like 'how do we hold this person accountable?' or 'was it human error?' or 'how close to achieving zero-harm are we?'.

Even if you've never heard about Safety Differently or Safety II, then this is still the book to read about safety in your organisation – whether you are a safety professional or not (and most likely, you're not). This little book gives you the best, shorthand and complete introduction to what you need to know and how it applies to you. You'll learn a lot, and you won't ever again look at safety the way you did before you started reading it.

Sidney Dekker
Professor at Griffith University in Brisbane, Australia, and at the Delft University of Technology in the Netherlands
Brisbane/Delft, Spring 2020

Preface

Improving safety and minimising incidents and accidents is the right thing to do. As a priority, it is right up there with making money, creating stakeholder value, saving the planet and maintaining a healthy work–life balance.

Over the last few decades, the performance goals that organisations strive after have increased tremendously in number. These objectives are changing rapidly and are often in conflict with each other. To manage and counter this complexity, companies have added structures, processes, rules and roles to their organisational bureaucracy. This is particularly visible in the safety domain where the number of rules that need to be adhered to is growing at an exponential rate, making it more difficult than ever before to comply with them all, all the time.[1]

Operational performance – including safety – relies heavily on your people's ability to navigate this complexity and adapt their behaviour as circumstances dictate. And yes, that does mean that they might circumvent the rules every now and then in your organisation's interest – very often with a successful outcome as a result. Only every so often does this behaviour lead to an adverse outcome, which you then hear about. In the aftermath of such an event, there will be real victims, people that you will need to talk to and explain why it wasn't possible to prevent their suffering. Higher management, regulators, press and politics will be all over you and demand answers that you can't easily give.[2]

In this book for the first time, the multiple small steps that collectively facilitate the improvement of safety – and even overall organisational performance – are described as tested in various industries across the globe. The approach that is outlined in this book is based on concepts that have been developed since the nineties by Sidney Dekker and many others. Following on from a career in industry, I was fortunate enough to collaborate with Sidney in implementing these ideas at various organisations over the last decade, discovering what works and what needed to be tuned to create acceptance and efficacy. I am very much indebted to Sidney and his affiliates for introducing me to these principles and allowing me to contribute to their development. I am also grateful to the organisations that have allowed us to observe their operations and identify weaknesses. These organisations have been brave enough to welcome our comments and suggestions in an effort to improve their operations. We applaud their candour.

In this work, I have tried to get away from specific paradigms, instead writing in layman's terms how to take small steps towards safety. I have tried to avoid safety jargon because that mystifies rather than clarifies and tends to create outgroups. Still, I have added a glossary in case you want to flaunt your expertise. Practical examples are included, and you will find supporting material referenced in the expansive notes section. These will prove particularly helpful for both the new-view sceptics and the safety science nerds that are interested in more detail, proof and further reading. I have ended each chapter with the key action points. Safety is not a fun topic, and I am hardly a humorous guy, but nevertheless I hope that this book is not only useful but also somewhat enjoyable.[3]

Notes

1. ... Over the last few decades...: The BCG Institute for Organization found that the performance requirements for business have increased sixfold between 1955 and 2010: from between four and seven to between twenty-five and forty. Morieux and Tollman (2014, p. 195).

   ... companies have added...: The BCG Institute for Organization found that the bureaucracy had increased thirty-five times between 1955 and 2010. This is visible, for example, in the growth of the number of meetings that people attend, the value of these and the number of reports to be written.

   Morieux and Tollman (2014, p. 195)

   ... in the safety domain...: In a 2019 study on health rules imposed by other businesses (rather than regulations that are created on behalf of the government), the UK Health & Safety Executive found that more than a third of the respondents felt that 'there is no real link between what they have to do for health and safety and actually keeping employees safe'. Furthermore, '39% of businesses reported feeling that taking responsibility for health and safety just feels like more and more paperwork, with no obvious health and safety benefit'. Health and Safety Executive (2019)

2. ... Operational performance...: The approach dictated in this book aims to enable work to go well and to increase the number of acceptable outcomes – addressing safety, quality, productivity, reliability, employee satisfaction and profitability. It therefore transcends the safety domain although that remains our primary focus. Hollnagel (2019) introduced 'synesis' as a new term to represent 'the condition where practices are brought together to produce the intended outcomes in a way that satisfies more than one priority and possible [sic] reconciles multiple priorities and also combines or aligns multiple perspectives'. Synesis is needed as a replacement for 'Safety II' because 'the meaning of the word "safety" in Safety II has little to do with the traditional interpretation of safety'.

   ... circumvent the rules every now and then in your organisation's interest...: Laurence (2005); de Boer, Koncak, Habekotté and Van Hilten (2011); Sasangohar, Peres, Williams, Smith and Mannan (2018)

3. ... have been developed since the nineties by Sidney Dekker...: The approach includes various compatible elements that are sometimes collectively known as 'Safety Differently' but are also labeled as Safety II or Resilience. See the glossary for an exacter definition of these terms.

... and many others...: The documented approach is based on earlier work by James Reason, Jens Rasmussen, David Woods, Erik Hollnagel, Nancy Leveson and many others in the safety science domain. I have added supporting insights from business scientists such as Clayton Christensen, Amy Edmondson, Yves Morieux and Dave Snowden. The approach benefits vastly from connections with business, psychology and human resources literature and methods.

... I am hardly a humorous guy...: You will find that I switch between the plural ('we') and the singular ('I') first person pronoun in this book to reflect the collaborative nature of the design and application of the method versus the lonely chore of putting the thoughts to paper.

Author

Robert J. de Boer MSc PhD (1965) was trained as an aerospace engineer at Delft University of Technology. He majored in man-machine systems and graduated cum laude in 1988 on the thresholds of the vestibular organ. After 20 years' experience in line management and consulting (Unilever, A.T. Kearney, Fokker), he changed career to pursue his scientific interest in team collaboration and safety, culminating in a PhD (achieved in May 2012) at the Delft University of Technology. From 2009 to 2018, Robert was appointed as Professor of Aviation Engineering at the Amsterdam University of Applied Science (AUAS). In this role, he executed research in the field of safety leadership, Human Factors and Restorative Practice, not just in aviation but across many different industries. Robert is currently the Director of the Amsterdam Campus for Northumbria University, combining this with safety research and a role as Director of the Human Safety Academy in the Netherlands. Robert has supported the improvement of safety at many organisations across the globe such as OKG (nuclear, Sweden), Mersey Care NHS Foundation Trust (health care, UK), Shell, BP, ExxonMobil, Neptune (oil & gas industry, global), Luton Airport, Lufthansa Technic, Thai Airways International, KLM, EASA, TUI, the Dutch armed forces, Dutch railways and the Dutch road authority.

1

Introduction

1.1 Introduction

> If only people would just abide by the rules, take care and be professional...

This book has been written to target those who are responsible for the overall performance of organisations, divisions or departments. Undoubtedly, your responsibility will include the need to maintain safety and avoid incidents. Most probably, you are not a safety specialist: you may have delegated the management of safety to a specialised department or professionals and entrusted them to help you maintain safety. Your safety officers seem to be well trained and competent. Nevertheless, every now and then, you may be unpleasantly surprised by incidents, perhaps even as a result of human behaviour.

This book is aimed at helping you avoid incidents, or at least understand them if they do occur, with an approach that is different, doable and directed. This approach represents a new paradigm of safety that focuses on how to integrate the natural variability of human performance – and our ability to compensate for unpredictability elsewhere – into organisational systems, thereby ensuring successful outcomes. Incidents are a natural consequence of the complexity of the system, whose effect we can diminish if we know what's going on and are able to react appropriately. The approach described in this book is particularly suitable for organisations that are subject to a low incident rate and where the cause of adverse events is not a single, easily identifiable failure or malfunction.[1]

This approach is not a substitution for existing safety rules and safety management systems but builds on them. Although a (safety) management system can lay a foundation for useful policies and effective processes, any safety effort is wasted if it is not aimed at substituting chronic unease for bureaucratic compliance, and understanding for metrics. Risks are less visible than ever before because there are less incidents to extract hazards from and many are hidden in complex interactions and automation. The cause of most, if not all, incidents is an amalgamation of conditions and decisions that reflect the current complexity of businesses, rather than just a single, easily identifiable malfunction or error. The approach in this book helps you to identify pertinent risks and to decide on their significance more effectively than using traditional means alone. Procedures and rules are not discounted but kept up to date and useful. Safety astuteness is cultivated across the organisation while augmenting existing safety policies and systems.[2]

This book does not address leadership styles, skills or attitudes. There is more than enough out there for you to read if you wish. Rather, I have tried to offer specific actions for you to take in the interest of safety, regardless of where you stand as a leader.

Besides managers whose buy-in is conditional for the adoption of these novel approaches to safety, this book is also intended for regulators that oversee safety performance of industries, and procurement professionals that monitor safety achievements across supply chains. This book will help them to set expectations for the organisations they watch over while combatting excessive bureaucracy. Finally, of course, the safety professionals themselves can use this book to advance safety within their own organisation and help explain what they are up to.

1.2 A Different, Doable and Directed Approach to Safety

Different to traditional approaches to safety, this approach presupposes that because of the complexity of organisations and variation from the environment, procedures are not always up to date or appropriate. Deviations from the rules are viewed as a trigger to analyse the situation rather than as a straightforward non-compliance. We uphold our judgement and seek to understand how work really gets done, especially when conditions are inadequate, and resources are slim. We create a framework of rules together with the front line and then empower them to take responsibility for any exceptions. Rather than using 'human error' as a cause of an adverse event, we aim to explain what triggered this: we assume that the error is a symptom of underlying causes, such as system design, training, supervision or conflicting goals. We recognise that safety margins are continuously being eroded in the absence of adverse events. We believe that numerical goals lead to manipulations that frustrate transparency and understanding. We know that in the wake of an adverse event, pain, pride, payments and publicity make for a resentful conundrum, and therefore recommend an alternative to the traditional forms of 'Just Culture'.

This approach is *doable*. In this book, I explain how to improve and maintain safety by focusing on the steps that need to be taken, aiming for clarity, brevity and practical application. I start by helping you and your team to identify where safety might be jeopardised and go on to help address those issues. I indicate how narratives and sensemaking help identify what is going on. I explain how to innovate sensibly, and how to monitor the effect of improvements on safety and performance. I explain how to keep the discussion on risk alive and ensure that operators are competent to maintain performance even in dire circumstances. And should things go belly-up, I have included a chapter on what to do – both to maximise learning and to restore the feelings of hurt and pain that are inevitably associated with adverse events.

This approach requires your top-down *direction* because the thinking behind it is somewhat controversial. The approach rejects the term 'human error', challenges the established culture of absolute compliance, suggests that 'zero-harm' policies are counterproductive and requires the acceptance of accidents as part of working life. You need to direct your team to be curious about events without passing judgement, finding out about how work really gets done. Press, public, politics and even our peers generally require us to adhere to our own procedures, and we need to work on improving compliance, but there is no way that you will get to know about what is really going on if you don't accept, even appreciate, the 'bad news' of existing gaps between rules and reality. Your direction is required to support your managers and safety staff in embracing the recommended approach and getting them to applaud unwelcome messages.[3]

1.3 Why You Might Want to Read This Book

If you are like many managers that we have met, you will believe that your people generally respect the rules and regulations that you have set for them. After all, these procedures have been well thought out and keenly deliberated, and are there for the good of the employees and the company. It is important that people stick to the rules and you take every opportunity to communicate that. Only every now and then, when something bad happens, it turns out that people have not adhered to what they are supposed to do – no wonder things turned out badly.

In actual fact, we have found in many organisations that there is a gap between the rules and how the work is actually executed, not just when things go haywire but most of the time. Because of conflicting goals, missing tools or a lack of resources, people have to make do. We just don't hear about it, because there is no reason to tell us and there is no reason to delve. The organisation is rewarded for the fact that the job gets done and at the same time we can maintain an aura of compliance, until the proverbial dreck hits the fan:[4]

> After a major leakage of twenty-nine thousand litres natural gas condensate into a canal near the town of Farmsum in 2018, the Dutch regulator for the oil & gas industry noted: "A major cause of the incident was due to the safety culture at the tank park. [...] Management has insufficiently succeeded in creating working conditions in which incidents are prevented: a culture in which sufficient critical attention is paid to

process deviations. There is too much confidence with management about the proper functioning of operational systems and procedures. [...] The [company] must draw up a program that aims to change this culture. This program should pay attention to the attitude of management (in particular promoting constant alertness, or 'chronic unease'), and on creating the working conditions that make this attitude possible (i.e. making available sufficient manpower to critically assess process deviations, creating internal debates, and promoting managers based on their critical thinking power, etc.)." The regulator notes that similar comments had been made as a result of a previous study several years earlier.[5]

This case was judged so reprehensible that a criminal investigation has been initiated against the company. In other cases, managers have been held *personally* accountable for safety breaches, and even incarcerated:

In 2016, following a fire at the Chemie-Pack company in Moerdijk five years earlier, the director, the safety coordinator and the production manager were sentenced to community service of between 162 and 216 hours as well as suspended prison sentences of twice 6 and 4 months respectively. The director and the safety coordinator were also banned from performing their job for a similar company for two years. The sentences would have been non-probationary if the case hadn't dragged on for so long and created so much publicity. The accident prevention policy had not been regularly reassessed and revised as should have been the case for a company handling hazardous goods. The appeals court held the managers personally accountable for this.[6]

Similarly, in the UK in 2016:

A director of a shop selling fireworks was sentenced to ten years imprisonment after a fire killed an employee and a customer. The court found that fireworks were prohibited from being stored in the shop, that the volume of fireworks exceeded what was permitted, and that the fireworks were stored too close together.[7]

These cases are examples of several recent verdicts where health and safety laws have been trespassed, resulting not only in fines for organisations, but also in custodial sentences for individuals in management positions in Australia, the UK, the Netherlands and Ireland.[8]

The approach documented in this book addresses the improvement points that are mentioned by the regulator in the case of Farmsum: creating working conditions in which incidents are prevented, paying attention to process deviations, verifying the proper functioning of operational systems and procedures, promoting 'chronic unease', making sufficient manpower available to critically assess process deviations, creating internal debates and promoting managers based on their critical-thinking power. Furthermore, our approach has been shown to reassure the regulator.[9]

This approach might even make you money:

Mersey Care, an NHS community and mental health trust in the Liverpool region, fundamentally changed its responses to incidents, patient harm, and complaints against staff according to the approach that I present in the chapter on restoring relations. The launch of this approach resulted in many qualitative improvements for staff, such as a reduction in suspensions and dismissals, increase in the reporting of adverse events, increase in the number of staff that feel encouraged to seek support and an improvement in the trend in sick leave. It also improved staff retention. Through interviews with Mersey Care staff and the collection of data pertaining to costs, suspensions, and absenteeism, the economic benefits of this approach were calculated. We found a significant reduction in salary costs, legal fees and termination costs, totalling approximately 2% of the annual total costs and 4% of the yearly labour costs.[10]

Despite our reluctance to express safety in metrics, this approach contributes to improved performance indicators:

Retail chain Woolworth's in Australia was persuaded to participate in the first deliberate, randomized control safe-to-fail small scale experiments in safety (see the chapter on effective workplace innovations). Staff members and managers of randomly selected stores were empowered to remove safety processes, procedures, checklists and rules that they didn't think were useful (except if specifically required by state or federal law), and were trained to identify and support each other's positive abilities that make things go right. The result was a significant reduction in the number of lost-time injuries and the award of Woolworth's annual safety prize.[11]

In an attempt to create safer working environments after an incident, many organisations have focussed on the deviations from procedures and implemented increasing layers of bureaucracy to prevent unwanted performance – but with limited success. This book offers an alternative, addressing safety continuously and coherently, not just intermittently in reaction to incidents. We will show how you can embrace non-compliance as a starting point to gaining insight in performance, rather than something to hunt after. We offer a new safety paradigm that will facilitate comprehension and learning – not just after incidents but continuously. We explain how to make sense of complex situations, avoid suspicion and deceit, and decrease the number of deviations from rules and procedures. It is not always an easy paradigm to appreciate and apply, but if adopted successfully, it will help you to understand and improve operational processes, it will increase the motivation of your staff, and best of all, it will support the organisation to function successfully under varying conditions. It enables the number of intended and acceptable outcomes to be as high as possible. In all, the new paradigm has been shown to enhance performance – not just in the safety domain but across the whole organisation.[12]

1.4 Reading Guide

The current chapter forms the introduction to this book. The next, second chapter is about what *you* might want to do as a leader to enable the new safety paradigm to be introduced. I describe how you need to welcome 'bad news' about deviations between how work is actually done compared to what the rules state, so that you can start closing this gap. When confronted by an adverse event or a significant gap, you might want to look more broadly than just focusing on those who were directly involved. I explain how you can get a better understanding of what is driving performance through combing points of view rather than relying on metrics. I urge you to refrain from adopting the traditional approach to 'Just Culture', aiming to restore relations and maximise learning instead. Finally, I explain what you should and should not expect from the safety professionals that are supporting you in this endeavour.

The third and subsequent chapters describe what *others* might do under your direction. In chapter three, I present what we intend to achieve with rules and procedures. I show why actual work might (and often does) diverge from this: rules and procedures do not take into account conflicting goals, are focused on compliance and not necessarily on getting the job done, do not take changing circumstances into account or wrongly assume that people are fully trained. By far most of the times that work is executed, the result is acceptable, and only every so often is the outcome unwanted. One thing we can study is the size of the gap between how work is executed and how it is prescribed in rules and procedures. When this gap is large, the goals that we strive for with rules and procedures do not get realised. In this third chapter, I present ways to see this gap, close the gap and make sure that we keep it closed.

In Chapter 4, the steps that can be taken to safely implement performance improvements are introduced. Ideas to better align actual work with the rules, for innovation, or when adapting to changing external circumstances need to be tested before they can be implemented. Micro-experiments are small initiatives to try out these improvements around a particular issue in a safe-to-fail manner, meaning that if the intervention is unsuccessful, it does not lead to disproportional damage. Micro-experiments are essential tools in a complex environment where outcomes cannot be predicted beforehand. I explain how to design and execute micro-experiments and give some practical examples.

After a period without serious incidents, many organisations are inclined to falsely assume that safety has been achieved. They ignore tell-tale signs of eroding safety margins that might be visible ahead of serious occurrences (as discussed in the fifth chapter). These include the pressures of scarcity, decrementalism, sensitivity to initial conditions, unruly technology and the contribution of protective structures. I then discuss what can be done to counter these pressures and maintain safety margins: keeping the discussion on risk alive and allowing operators to gain experience at the edges of the operating envelope.

Chapter 6 is devoted to what we might do if – despite all our efforts – things go wrong after all. Even though the aftermath of an accident is often a melting pot of emotions, blame, litigation and hurt, there are ways to ensure that relations are restored and that the maximum effect of learnings is available to the organisation. I also discuss ways to ensure we receive reports about safety issues.

I end this book with the encouragement to apply these recommendations in practice. In Chapter 7, I also discuss how different industries each have achieved a different level of safety, and how this might require a different emphasis in our approach. I hope that this chapter is not the end of your journey, but only just the start.

Notes

1. ... ensuring successful outcomes...: Hollnagel, Wears and Braithwaite (2015).
2. ... but builds on them...: 'builds' is used here in the figurative sense. It is probably not necessary to increase the number of safety policies and systems, but to use what exists and make it more effective. In fact, 'decluttering' safety rules and policies (pushing back on bureaucracy, and compliance for the sake of it) is usually a result of applying this approach. Dekker (2017, p. 193).

   ... existing safety rules and safety management systems...: A Safety Management System is a systematic approach to managing safety, including the necessary organisational structures, accountabilities, policies and procedures. In domains like aviation, such a system is required for many companies in an industry-wide effort to improve safety even further. A vital component of contemporary safety management systems is risk identification. But there is little that you can do to manage and mitigate a hazard that you don't know about. And the traditional approach is fraught with challenges; see Anand (2015a, 2015b). Our approach helps you identify these risks and generate safety astuteness across the organisation. See, for instance, https://www.skybrary.aero/index.php/Safety_Management_System, accessed August 23rd, 2018. ICAO (2013a, 2013b).

   ... The cause of most if not all...: Take as an example the Deepwater Horizon / Macondo accident: the BP investigation report lists eight technical failures (safety breaches), suggesting that these are the main causes of the accident. However, they also suggest elsewhere in the report that: '...a complex and interlinked series of mechanical failures, human judgements, engineering design, operational implementation and team interfaces came together to allow the initiation and escalation of the accident. Multiple companies, work teams and circumstances were involved over time'. Dekker (2019, p. 320).

... that reflect the current complexity of businesses...: The BCG Institute for Organization found that the performance requirements for business has increased six-fold between 1955 and 2010: from between four and seven to between twenty-five and forty. Morieux and Tollman (2014, p. 195).

... identify pertinent risks and to decide on their significance...: for a discussion on the weaknesses of traditional risk assessment methods. Bakx and Richardson (2013).

... Procedures and rules...: In a 2019 study on health rules imposed by other businesses (rather than regulations that are created on behalf of the government), the UK Health & Safety Executive found that more than a third of the respondents felt that 'there is no real link between what they have to do for health and safety and actually keeping employees safe'. Furthermore, '39% of businesses reported feeling that taking responsibility for health and safety just feels like more and more paperwork, with no obvious health and safety benefit'. Health and Safety Executive (2019).

3. ... improving compliance...: I hesitate to use these words, because as will become clear in later chapters, compliance is a step that comes late in this approach. We first need to understand non-compliance.

4. ... not just when things go haywire...: In one recent case it was suggested that specific non-compliances leading up to an adverse event were acts of sabotage, completely missing the point that non-compliances are very common and generally turn out satisfactory.

5. NRC.nl, October 9th, 2018: Strafrechtelijk onderzoek naar NAM na lekken gif in Farmsum. NAM (2018): Staatstoezicht op de Mijnen (2019); letter to Nederlandse Aardolie Maatschappij B.V. dated May 27th, 2019, reference number 19131749.

6. High court of Den Bosch, April 28th, 2016: verdict in case 20-004376-12 / ECLI:NL:GHSHE:2016:1594.

High court of Den Bosch, April 28th, 2016: verdict in case 20-004378-12 / ECLI:NL:GHSHE:2016:1596.

High court of Den Bosch, April 28th, 2016: verdict in case 20-004379-12 / ECLI:NL:GHSHE:2016:1597.

7. Brodies. (2018). Custodial sentences for health and safety breaches. https://brodies.com/binformed/legal-updates/custodial-sentences-for-health-and-safety-breaches?utm_source=Mondaq&utm_medium=syndication&utm_campaign=View-Original, accessed July 21st, 2019.

8. ... in Australia...: In Australia, the alleged first custodial sentence for a safety breach was imposed late 2018. After a 2017 workplace fatality of an employee, the owner of a scrap metal business was convicted and sentenced to 6 months' imprisonment and a fine of AUD 10,000. The penalties are still subject to appeal.

Helmore, S. (January 16th, 2019). First prison sentence handed down for breach of WHS duty offence. https://www.sparke.com.au/insights/first-prison-sentence-handed-down-for-breach-of-whs-duty-offence/, accessed July 21st, 2019.

... the UK...: In 2016 in the UK, a care home director pleaded guilty to the manslaughter of a resident and was sentenced to 3 years and 2 months' imprisonment. And again in 2016, a director of a construction company that was dismantling a building was sentenced to 6 years' imprisonment after an employee died on the site. A year later, another director of a construction company was jailed for 14 months in July 2017 for the deaths of two workmen during the refurbishment of a London flat. In June 2018, a husband and wife were each sentenced to three years' imprisonment for manslaughter by gross negligence, after a girl died when a bouncy castle was carried away by the wind. Brodies (2018).

… and Ireland…: A company director was jailed for four years in January 2020 after a warehouse employee was fatally crushed by planes of glass. The judge justified the verdict by stating that the director did not provide a risk assessment or supervision for a task that required both. 'He just told them to get on with it', the court was told. IOSH. (2020). Company Director jailed for four years. *IOSH Magazine*, March / April 2020, p. 23, available at https://issuu.com/redactive/docs/iosh_marapr2020_full_lr_35ef90d10d1084, accessed April 18th, 2020.

9. … The approach documented in this book…: Note that our approach is not directly aimed at a reduction in the number of adverse events (incidents & accidents). The reason is that safety in many industries can no longer be considered as an absence of accidents and incidents because the frequency of these is (fortunately) quite low. Safety can't be counted. Rather, safety is considered as a system's ability to function as required under varying conditions.

… has been shown…: An oil & gas production company had a tainted relationship with the regulator following several incidents and the lack of compliance to procedures that became visible during the ensuing investigations. As part of the turn-around, work instructions were improved in collaboration with the operators to better highlight the critical points and reflect how work was done in practice. The regulator reacted very favourably, and this helped rebuild trust.

10. Kaur, de Boer, Oates, Rafferty and Dekker (2019).

11. Dekker, S. (2018). The Woolworth experiment. http://www.safetydifferently.com/the-woolworths-experiment/, accessed November 2nd, 2018.

12. … increasing layers of bureaucracy…: The BCG Institute for Organization found that the bureaucracy had increased thirty-five times between 1955 and 2010. This is visible, for example, in the growth of the number of meetings that people attend, the value of these and the number of reports to be written. Morieux and Tollman (2014, p. 195).

… shown to enhance performance…: This is the central tenet of this book and will be illustrated separately for each component of the approach.

2

Your Role as a Leader

2.1 Introduction

In this chapter, I describe how you can create the conditions to enable the new approach to safety to be successfully introduced in your organisation. This chapter describes what *you* might want to do as a leader to enable this new safety paradigm. In later chapters, I will then go on to describe what *others* might do under your direction. Here I describe how your top-down direction is required to support your managers and safety staff in adopting the approach that I present in this book. I will discuss how you need to welcome 'bad news' about gaps between how work is actually done and what the rules state. When confronted by a significant gap or an adverse event, I urge you to refrain from deciding which penalty to invoke using a traditional approach to 'Just Culture', and instead aim to restore relations and maximise learning. I go on to explain what you can do to get a better understanding of what is driving performance, and how safety targets can limit the transparency in your organisation. They actually drive incidents underground. I demonstrate how pursuing a safety culture is a roundabout way of addressing safety. Finally, I explain what you should and should not expect from the safety professionals that are supporting you in this endeavour.

2.2 Welcoming 'Bad News'

As in any organisation, your people are aiming to make progress in specific circumstances and need to choose their resources accordingly. This might mean that they are not always complying with the rules and regulations that have been set, or that they make errors, as they strive to achieve multiple goals. Your role as a safety leader requires you to promote a sense of candour and allow people to share mistakes and concerns without the fear of retribution. You need to welcome 'bad news' (i.e. information about gaps between rules and practice, or the fact that a near-miss occurred) as a chance to learn, not as something that damages your reputation. Having large gaps between rules and reality is bad – but not hearing about them is far worse:[1]

> The management of an aviation maintenance company felt it was plagued by too many incidents. Each time an incident occurred it turned out that the procedures were not adhered to and (as management felt it) shortcuts were taken. Management had asked for suggestions from the mechanics to improve the situation, but none were forthcoming. We were brought in as external consultants and allowed to talk to a group of mechanics without their supervisors and were quickly confronted with multiple examples of gaps between rules and procedures and how work was actually done. However, the mechanics did not allow us to share these with management until a long list of assurances had been given, such as: no anger and immediate scolding when confronted with cases of non-compliance; the examples that were brought forward will not affect progression or employment; incident investigators to be trained in understanding the gap between actual work and how rules are drawn up; anything that the colleagues shared would not be used against them; and all shared examples would be discussed and studied to improve the standards and quality of the procedures.[2]

This example shows how the truth is concealed if employees sense that it is not acceptable for management to hear about how work really gets done. The challenge for many managers is to be curious without being normative. Many of us find it extremely difficult to discuss how work is actually done without referring back to how we imagine the task 'should' have been done. Managers are continuously using rules and procedures as the baseline, conducting what might be perceived by the front-line workers as an inquisition rather than a dialogue. We have heard operators tell us that 'Safety rules are meant to protect assets and [the company], not the front-line workers'. There is no way that you will get to know what is really going on if you don't accept, even appreciate 'bad news'. No employee in their right mind will tell you about gaps between policies and how they actually do their work if they feel that it will negatively impact their job security or their relationship with their co-workers. Why bother - even without those threats? We need to not just objectively appreciate the bringing of (what is often conceived as) bad news, but also ensure that there are no perceptions that the opposite is true. This is particularly difficult as press, public, politics and even our peers generally do not understand why we would not adhere to our own procedures. One way to ensure that your direct reports are listening properly to their people is to expect them to tell you about any concerns in their area before you hear about it in any other way.[3]

2.2.1 Creating Psychological Safety

As a leader, you need to create an environment that is conducive to honesty and transparency, allowing bad news to be heard and acted upon. The term 'psychological safety' is used to describe an environment in which people feel that they can make and own up to mistakes, where people are able to bring up problems and tough issues, where people are not rejected for being different, where people feel supported and where people can ask for help. People even feel encouraged to innovate and experiment within certain bounds. We need to welcome differences of opinion and confirm that these do not jeopardise advancement opportunities or job security. In fact, we must convey that conformity in judgement is deadly. Psychological safety is not (just) about getting shy people to speak up: it affects everybody in the organisation, not just certain individuals. Psychological safety is sometimes mistaken for 'just being nice', but the opposite is actually true: we can be candid, have discussions and even fierce disagreements, as long as we acknowledge these are work-related and we retain mutual respect. Psychological safety allows us to improve our performance because it allows us to discuss and set standards and then hold teams and people accountable for these. We can improve our organisation's psychological safety by emphasising what we want to achieve and why it matters, but also by setting (realistic) expectations about failure and destigmatising it. Other tactics to improve psychological safety are inviting participation by demonstrating humility, asking good questions and listening intensely. We can create forums for input and provide guidelines for discussion; we need to express appreciation when input is delivered.[4]

Our employees need to feel (psychologically) safe before they will be honest with us and share their experiences with us:

> After the management of the same aviation maintenance company had given the requested assurances, we started organising further discussions with the team of mechanics. A trusted quality assurance officer led bi-weekly meetings on our behalf. Further examples of gaps between paper and practice were considered, and improvement suggestions were collected. These were documented by the quality assurance officer and reported to us (the external consultants). Of course, senior management was also keen to know what was coming out of these meetings – they felt that they were accountable for any existing gaps between Work-as-Done and Work-as-Imagined. But we did not pressurize the group as trust was starting to build, allowing them to decide for themselves if – and when – they were ready to start the dialogue with management. We had to remind management that gaps had existed for quite some time, whether they knew about them or not. If there had been no immediate urgency to close them earlier, why rush now at the expense of transparency? All-in-all it took several months before the mechanics were ready to speak up, and even then only through one specific and trusted manager.[5]

Different types of psychological safety can be discerned: whether new members are accepted into a team, whether it is appropriate to make mistakes and learn, at what stage someone is considered competent enough to act and contribute autonomously, and whether challenges to the status quo are tolerated without personal repercussions. Particularly, the latter is thought to promote creativity within the organisation.[6]

The most important skill to understand what is going on in your organisation is to encourage employees to voice their concerns and ideas with which they intend to improve the organisation – even if others disagree on these ideas. Speaking up is threatening for the recipient, independent of the content, because it challenges authority. It disrupts daily activities and adds workload. Employees can choose to remain silent when they expect a negative reaction, even if they have important information to share, because it involves a personal risk. Not surprisingly, the perceived openness of a leader to speaking up is strongly correlated with concerns being voiced. Leaders are sensitive to how a message is framed (as an opportunity or solution, or as a problem or threat) and how its urgency is conveyed, which might actually camouflage the importance of the message. Leaders are able to impede the articulation of concerns by becoming defensive or even angry, by setting ground rules about how and when to voice concerns, by avoiding an issue, or by disregarding or downplaying a trend and suggesting that the concern describes a rare circumstance. You will find that the people around you will mirror how you react to any concerns that are raised.[7]

What to do if on your rounds you see a large deviation from the rules – how do you react? Speaking for myself, I will not let the act go unaddressed if I feel that the action is not aligned with my own principles. Introducing myself (if needed) and explaining my surprise feel like a good start. But then I need to refrain from judgement and give ample room for my colleague to explain why the action makes sense to him or her, so that I really understand the rationale. In the process, I should be able to discover where the lack of resources and tools, the constraints surrounding the task and the conflict of goals all worked together to make the course of action coherent. I might want to make the worker's manager aware of the way that work is actually done, but for that I need to be prudent for his or her immediate reaction. The response might be more normative than I would wish, and so I would refrain from mentioning names but rather invite the manager to walk around and find out for himself/herself how work is done within his area of responsibility.

2.2.2 Avoiding Retribution

An adverse event (an accident, but also a significant near-miss or even a large deviation from procedures) generally creates a lot of hurt, reputational damage and loss of trust. Of course, the immediate pain is felt by the direct victims (if any). Generally, there will also be practitioners involved who feel personally responsible and remorseful for the event and its aftermath. Similarly, the organisation (represented by management) will feel blemished in its reputation and the general public may be distraught by the events. Severe incidents tend to generate a lot of attention, also from groups that consider all incidents and accidents to be avoidable – which inevitably suggests that there must be someone to blame for them. This is hardly a helpful paradigm from which to learn and improve, yet reflected in the traditional 'Just Culture' approach that is manifest in flow charts at many organisations.

A traditional 'Just Culture' is aimed at differentiating between acceptable and unacceptable performance and using this to determine appropriate consequences. A traditional 'Just Culture' is more focused on *who* is responsible than learning about the factors that led to the event, potentially leaving these unaddressed. A traditional 'Just Culture' approach leads to a lack of reports, honesty, openness, learning and prevention. A traditional 'Just Culture' tends to consider the existing procedures and rules as legitimate even though the rules may not have been suitable in the context of the event. This is exacerbated if the rule makers or rule owners are involved in the evaluation of the 'offenders'. And whereas a traditional 'Just Culture' in organisations seems to mirror justice in society at large, the crucial elements of an independent judge and the right to appeal do not exist. Not coincidentally, managers believe their organisation to be more just than those lower in the hierarchy.[8]

In this book, I propose a different approach to adverse events that avoids retribution. It relies on the recognition by the practitioners of the reputational and physical damage that is done to the organisation, and your acknowledgement of the pain and remorse that those that contributed to an adverse event feel. This process (called 'Restorative Practice') is described in detail in Section 6.5. It has been shown to reduce blame, enable transparency, improve motivation, facilitate learning and create economic benefits. A restorative approach takes time and is effortful, and so it does require your continued support to enable it to generate these welcome results.[9]

Taking a restorative approach implies that a single deviation from procedures, even with grave consequences like an accident, should never be a reason for punishment (at least within organisations; public prosecutors may assess this differently). Even if it transpires after an incident that the violation was limited to one specific individual and occurred repeatedly, I still strongly recommend not to use the event as the trigger for disciplinary action. Rather, this suggests the need for a discussion as to why the deviations went unnoticed to peers and management for so long, and what additional training and supervision is required for the individual. Only after these have been attempted does it make sense to perhaps reconsider whether the job fits the person. But if a change of roles is being considered, then this must become apparent during normal work, and definitely not follow as a result of an adverse outcome because otherwise this will jeopardise transparency and the commitment of co-workers.

So if adverse events are a natural consequence of the complexity of the organisation in its environment, why do corporate presidents, the press and politicians keep blaming individuals and calling for retribution when things go wrong? According to some researchers, blaming others helps to distance oneself from the accident and to absolve oneself of guilt. This is not only valid for individuals but also for complete departments or organisations, which can avoid further (possibly damaging) investigations by directing attention to an isolated human failure. The thought that some maverick is responsible for the adverse outcome further maintains the illusion of control in an orderly, rational world – rather than accepting that things are uncertain, and disaster can therefore strike again without warning. In our modern (western) society, we are not accustomed to a lack of safety and reject discomforts as a result of black-outs, natural disasters or just plain bad luck. In these cases, we believe – even expect – that there must be someone to blame that either caused the inconvenience or failed to prevent it. It is your role as a leader to counter these misconceptions because they stand in the way of transparency, safety and performance improvements.[10]

2.3 Setting the Scene

In this section, I go on to explain what you can do to get a better understanding of what is driving safety performance in your organisation, so that the organisation is able to start making amends as described in Chapter 3 and the subsequent chapters of this book. I discuss how to make sense of complex situations that give rise to (a lack of) safety. I discuss the disadvantages of using safety metrics as targets and propose that you disregard safety culture as something to test or tweak.

2.3.1 Making Sense of the Situation

> Road inspectors recently saw and photographed, see figure 2.1, a road worker crossing a freeway twice amidst traffic speeding along at 100 kilometres an hour (62 mph) to reinstate a speed limit sign after the road works had been completed. Asked about this illegal and seemingly frivolous behaviour by the inspectors, it turned out that the road worker had asked his supervisor for permission to risk his own life amidst the oncoming cars. The road worker and his boss agreed it would be far better to quickly complete the job (despite the need to twice cross the two-lane freeway) rather than to wait several hours for the safety car to arrive that would protect anyone crossing. These safety cars should have been available – if it hadn't been for some planning errors that were made in the weeks leading up to the job.[11]

This example demonstrates how conflicting goals (time pressure versus higher personal and reputational risk) and lack of resources create a gap between rules and how work is actually done. It also vividly shows that the road worker's dilemma is caused by people (the planning department) that are detached in time and location from the reality of the moment, and how the subsequent solution is endorsed by his boss. In this case, as in real life generally, decisions are made with the goal to take action and achieve some kind of result.

FIGURE 2.1
A road worker crossing a freeway amidst traffic speeding along at 100 kilometres an hour (62 mph) to reinstate a speed limit sign after the road works had been completed. (Photo by Rijkswaterstaat. Used with permission.)

Decision-making is part of the process of understanding the situation, linking this to earlier experiences, trying different courses of action and evaluating the results. Rather than weighing all alternatives beforehand, professionals tend to try a course of action that worked previously in (what is perceived as) similar circumstances and modify this as they evaluate the effects. This approach is particularly common (and energetically efficient) for every day operational issues. We tend to favour autonomous, subconscious cognitive processes as long as possible, only switching to the more effortful conscious thought process when cues from reality become so overbearing that we really can't put it off any longer. Of course, every now and then what seemed to be an acceptable outcome turns sour, and we wish we had done more thinking.[12]

The road authority was initially shocked at the behaviour described above, but they were able to supress their first inclination to reprimand the road worker and sanction his boss. They only came to know about the considerations of the road worker and his boss because they engaged with them and were inquisitive. They did not hide their surprise and their rejection of how the situation evolved but were genuinely interested to know why the actions made sense to the road worker and his boss at the time. As a result of this case, new arrangements were made with the planning department about how to prepare for road works. It was also made clear to the road worker and his boss that repetition of this behaviour is not acceptable for a company wishing to continue working for the road authority.[13]

To understand what is happening in a complex environment, we need to engage in *sensemaking* – a process that allows a confusing situation to be discussed and comprehended explicitly in words and that serves as a springboard for action. Each involved party tells their story or account of how the situation was experienced. Working together, the event can be reconstructed from all angles, and a coherent reality created. To make sense of safety-critical situations and events like the one described above, we need to engage with all those involved. In the case of the road worker, this includes at least the road worker himself, his co-worker on-site, the supervisor, the planning department and the safety inspectors. In particular, the exchanges between stakeholders need to become apparent because these interactions often drive unanticipated behaviour. Opportunities for improvement emerge from the resulting description of the event, allowing decisions to be taken and implemented. Rather than driving for quick fixes, it is your role as a leader to ensure that you allow sensemaking to take place and that it is followed up appropriately.[14]

2.3.2 Trading Targets for Transparency

Safety targets have a detrimental effect on transparency, honesty and cooperation. Safety targets are usually aimed at minimising minor injuries in the misguided assumption that this reduces the probability of fatalities or a major accident. Back in the 20s of the previous century, a safety engineer for an insurance company identified from the company's records that for every accident with major injuries there were 29 accidents with minor injuries and 300 accidents without any injuries at all. This is often misinterpreted as a 'law of nature' that is used as an excuse to target trivial safety violations in an effort to improve overall safety. Famous is the photo in the *New York Times* (Figure 2.2)

FIGURE 2.2
Randall Clements, left, plant manager of DuPont facility in LaPorte and DuPont spokesman Aaron Woods, right, walk out of the plant to speak to the media about a gas release that killed four employees on Saturday, November 15th, 2014. (Photo by Marie D. De Jesus/© Houston Chronicle. Used with permission.)

where two Dupont managers are on their way to speak to the media after four workers died following a hazardous gas leak – passing a sign to 'please take extra precautions when driving and walking' (a seemingly petty safety hazard under the circumstances). Just before the disaster with the Deepwater Horizon offshore drilling unit in the Gulf of Mexico in 2010, BP awarded prizes to that very rig for an impressive reduction in the number of occupational accidents.[15]

Several recent studies have confirmed the *negative* correlation between fatalities and injuries in many industries such as oil and gas, aviation and the construction industry. These studies show that *more* (recorded) injuries lead to *less* fatalities. If this seems counter-intuitive, consider how important it is to promote transparency, and to avoid any downward pressure on identifying and rectifying dangerous situations. It is possible to hide an injury, but you can't hide a corpse. The use of safety metrics like Lost Time Injuries will lead to risk secrecy and a disconnect between layers within a hierarchy, typically eliminating any opportunity for learning:[16]

> We were on an gas platform in the North Sea when we heard about a medical case that had occurred during the previous night. Apparently, the eye of a subcontractor's rigger had become irritated, perhaps by some dust particle that was flying around as they were sanding the rig at the time. He was adamant however that there was nothing in his eye and that he woke the nurse as a precaution only – because otherwise it would be recorded as a medical incident and disadvantage his employer. In fact the employer was already on the telephone from shore in an effort to ensure that the incident would not count against his supplier performance metrics while the rigger was still with the nurse.[17]

Similarly, companies in the construction industry in the UK that pursued zero-harm targets had a *higher* number of fatalities than those without such policies – a surprising and worrying finding given the ubiquitous adoption of zero-harm targets in many industries.[18]

Many companies struggle to find a balance between productivity and safety. Despite the slogans and banners, safety is not the only objective of any organisation. To suggest it is disregards (even disrespects) the choices your people are making every day out there in the field. It is your role as a leader to support people in balancing safety with other targets, and to ensure that safety is considered (an important) part of overall performance. I urge you to eradicate stand-alone safety targets and avoid their negative effect on transparency and psychological safety. Rather, replace bureaucratic compliance and the use of one-dimensional metrics with chronic unease and sensemaking.[19]

2.3.3 Circumventing Confusion about Culture

In the case of the gas condensate leak in Farmsum described in Section 1.3, the regulator required the company to improve its 'safety culture'. By this they meant that: working conditions were created in which incidents are prevented, attention is paid to process deviations, the proper functioning of operational systems and procedures is verified, 'chronic unease' is promoted, sufficient manpower is available to critically assess process deviations, internal debates are held, and managers are promoted based on their critical-thinking power. I suggest that all of these are desirable and even necessary, but that they need not be blanketed by the confusing term 'safety culture'. In this example, it will be hard if not impossible for the regulator to assess the proposed programme to improve the 'safety culture', and to evaluate the effectiveness of the programme after it has been implemented, unless the evaluation is based on an approach similar to what I describe in this book. Note that the regulator is certainly not asking the company to expand or harshen its safety rituals (holding bannisters, parking backwards, no phoning while driving) that are sometimes included under the umbrella of a 'safety culture' but only serve to distract from the real issues.[20]

Chesley B. 'Sully' Sullenberger landed his aircraft on the Hudson River after both engines malfunctioned in January 2009, saving everybody on board. Richard Champion de Crespigny landed a Qantas Airbus A380 safely in Singapore with 440 passengers and 29 crew after an uncontained engine failure severely damaged the aircraft on November 4th, 2010. Both men are typically acclaimed as heroes because they were able to avert a tragedy in dire circumstances by a combination of creativity and skill that seems literally extraordinary. However, these successes relied heavily on a wider system that includes designers, builders, maintainers, operators and rescuers working together. Just as we cannot blame individuals when things go wrong but rather look at the broader context, similarly when things go well, it seems arbitrary to congratulate only those that were directly involved. Do away with your safety awards. Acknowledge that the real heroes are often invisible; they are the ones that understand why things go right. They keep their processes functioning smoothly under varying conditions, so that the number of intended and acceptable outcomes is as high as possible.[21]

Like any culture, a safety culture is inherently stable (requiring multiple years for noticeable changes) and it is ineffective to tweak just single elements of a culture. Focusing on culture means that many years will pass before a change will be perceptible, and in the meantime, the results will be confounded by other internal and external influences. Ritually, a safety culture survey might be rolled out, but these assessments have very limited predictive value regarding incidents. Depending on the exact definition, an advantageous safety culture is just as much a consequence of operational improvements, as it can be the cause. For instance, indicators of both 'safety culture' and psychological safety are that staff (including novices) can speak up and discuss concerns openly. As a leader, you don't need to use ambiguous terms like 'safety culture' and 'heroes' or award prizes to improve safety. You don't need slogans and posters either – they are not effective, belittle your people and distract from the real issues. Instead, you can rely on diligently executing the consecutive small steps described in this book to help you on your way.[22]

2.4 Redirecting the Safety Department

Your natural inclination might be to assume that much of what is explained in this book is work for the safety department. And actually, you will find that many in your organisation might agree. Line managers get off the hook for actions and responsibility, and the safety staff can get their teeth into something. But you'll get perfunctory results: safety improvements need to be owned by the operational team, not some staff department, because otherwise things won't stand up to the duress of routine work. 'In particular, safety professionals often have a nasty habit of telling people how to do tasks that the safety professional has not, in fact, ever done'.[23]

There is, however, an important role to play for safety professionals: they can help roll out this approach, support and facilitate groups of workers, and monitor progress for you. They can support you and other safety leaders that are keen to follow the documented approach but feel inhibited because of external and internal factors such as a lack of resources or excessive documentation requirements. They can ensure that the gaps and risks that are being identified tie into the Safety Management System. They can maintain the integrity of the reporting system and ensure that legal obligations, industry standards and supply chain requirements are met. Once a gap between rules and practice has been closed, they can be asked to check upon exceptions, and they are often great in doing investigations whenever an incident has occurred. Safety professionals can help managers identify how work is actually done (as long as they leave their normative attitude at home), design and execute micro-experiments, and help apply a restorative methodology. They can signal the need for operators to gain experience at the edges of the operating envelope in a safe-to-fail manner and help to facilitate this. And they have a role to play in maintaining a discussion on risk, but I would not give them the final word.

2.5 Conclusion

As a leader, you create the conditions to successfully introduce the new approach to safety, and to support your managers and safety staff in effecting this. You need to welcome 'bad news' about gaps between what the rules prescribe and how work is actually done. I have urged you to refrain from adopting the traditional approach to 'Just Culture' and instead look more widely at what contributed to the event. I have explained how to make sense of a situation and urged you to avoid quantitative targets or ambiguous terms to safeguard transparency and honesty. In the last section, I explained how to ensure the safety department is supporting you in applying safety leadership.

Key actions:
- Create psychological safety by welcoming differences of opinion, destigmatising failure and allowing people to innovate and experiment.
- When confronted with bad news, count to ten to supress your anger. Then be inquisitive, not normative.
- Expect your direct reports to bring you bad news before you hear it from others.
- Refrain from applying a 'Just Culture' flow chart or seeking retribution after bad news.
- Push back on pressure from peers, superiors or press that demand punishment.
- Ensure that it makes logical sense to all those involved how an event occurred by allowing everyone to tell their version before you move forward. Don't stop at 'someone messed up'.
- Eliminate or at least limit the use of safety metrics insofar as this is allowable. Don't set safety targets, don't award prizes, and don't use posters and slogans.
- Don't get side-tracked by discussions about culture or leadership styles. Focus on what can be done here and now.
- Get the safety professionals on board with this approach, and enable them to support line management with their responsibility for safety.

Now that I have described what *you* might want to do as a leader to promote safety, I now turn to what *others* might do under your direction.

Notes

1. ... choose their resources accordingly...: In a recent study in the oil and gas industry, it was found that generally, the operators thought favourably of the procedures. Reasons for frustration and deviating from them included information overload, outdated procedures, and a disconnect between writers and users. Not the criticality of the task but the frequency of the task and worker experience determine the use and perceived importance of the procedures. Sasangohar, Peres, Williams, Smith and Mannan (2018).
    ... that they make errors...: Amalberti (2001).
2. Aviation maintenance company, own experience, May 2019.
3. ... We have heard operators tell us...: Oil & gas industry, own experience, 2018.
    ... no way that you will get to know...: Ioannou, Harris and Dahlstrom (2017)
    ... objectively appreciate...: There are at least five ways of looking at deviances from procedures, most of which are detrimental for really understanding what is going on. These are the 'bad apple' theory, control theory, subculture theory, learning theory and labelling theory. The application of each of these

theories can be evidenced in many organisations, even if the theories themselves are not explicitly acknowledged. I will describe each in turn to ensure you recognise and can eradicate them, because they all serve to conceal what is really going on. The *bad apple theory* proposes that rule breaking happens because of bad apples in an otherwise good basket. Most people are well intended but a few of them are error prone or even delinquent. This way of thinking naturally focuses on individual characteristics and ignores or underestimates systemic and environmental factors that are part of everyday work. Hidden complexities that distort performance figures and gross ignorance of the meaning of central tendency lead to damaging witch hunts that do nothing for the long-term safety. Sometimes training needs are identified after the fact, but hardly ever do we take into account recruitment, coaching, supervision and feedback. This way of understanding takes Work-as-Imagined for granted and makes us blind for the daily complexities that front-end workers experience. The bad apple theory is still very much in evidence, particularly in the health industry, as this example shows:

"A small number of doctors account for a very large number of complaints from patients: 3% of doctors generated 49% of complaints, and 1% of doctors accounted for 25% of all complaints. [...] Continued legitimate complaints from patients warrant restricted licenses or removal from practice." Shojania and Dixon-Woods (2013, p. 2).

Closely related to the 'bad apple' theory is *control theory*. This suggests that rule breaking comes from rational choice: lack of external control enables it. Application of this point of view suggests that more monitoring and effective feedback mechanisms (perhaps punishment?) are the preferred instruments to close the gap between Work-as-Done and Work-as-Imagined. Companies that apply this point of view might invest in exhaustive management information systems, camera surveillance and time clocks. They will have a comprehensive listing of possible disciplinary actions, maybe even described as 'Just Culture' complete with process flow. A premise of control theory is that there is one correct way of doing things. The *subculture theory* emphasises the discord between different groups within the organisation, in particular the tension between management and workforce. Rule breaking happens because the rules are considered invalid and stupid by some. Subcultures that defy the rules gain traction and status. Management retorts by interpreting rule breaches as misalignment with the desired company culture, reinforcing the chasm between subcultures. An example that we often encounter is when rules are not congruent with production goals and a 'can do' subculture is cultivated, in which those that 'get stuff done' with little regard for the rules are considered heroes. The subculture theory is reinforced by a lack of independence between rule makers and rule enforcers. Enforcers who side with management then have a stake in compliance because they themselves made the rules. *Learning theory* posits that rule breaking comes from learning and fine tuning of daily work. People learn the positive and negative consequences of breaking rules under conditions of limited resources and goal conflicts. People will find the shortcuts in any task to get the job done, often instructed by more experienced colleagues (talk about expedited learning!). A famous example is the case of the Columbia space shuttle but I'm sure there are many in your own organisation as well – visible by reliance on old hands to show the new guys the ropes. Administrative steps that take (too much) time are often a case in point. Learning theory assumes that the gap between Work-as-Done and Work-as-Imagined can't really be closed; we need to rely on craftsmanship to get stuff done. Management condones this and turns a blind eye; the 'can-do' mentality is spread throughout the organisation. *Labelling theory* goes even further and assumes that Work-as-Done is necessary and correct. It downplays the gap between Work-as-Done and Work-as-Imagined by suggesting that it is just a matter of words: rule breaking doesn't come from the action we call by that name – it comes from us calling it by that name. Professionals know how to get their job done under taxing circumstances and the procedures that we give them are just guidelines to apply at their discretion. We see this kind of theory being applied in organisations of highly educated professionals like hospitals and air traffic controllers. Dekker (2014b, 2019, pp. 391–429); Shojania and Dixon-Woods (2013).

   … expect your direct reports…: Morieux and Tollman (2014, p. 171).
4. … psychological safety…: Edmondson (2018).
5. Aviation maintenance company, own experience, August 2019.
6. Clark (2020).
7. … to voice their concerns…: 'Voice' (noun, as in voicing concern) is defined as a form of organisational behaviour that involves 'constructive change-oriented communication intended to improve the situation', even when others disagree. Voice serves as a 'seed corn for continuous improvement' and allows employees to channel their dissatisfaction with the status quo. Voice behaviour is focused on correcting mistakes, improving processes, and formulating solutions to organisational problems. Voice is generally threatening for the recipient, even if the content itself is not, because it challenges authority. The effect of 'promotive' voice (ideas about improvements, solutions and possibilities) and 'prohibitive' voice (expressions of concern and harmful factors) on the leader's reaction differs. Sharygina-Rusthoven (2019, pp. 44, 61, 68, 93, 109).

   … Leaders are able to impede the articulation of concerns…: Leaders may not be aware that they enjoy 'leaders' privilege' that enables them (more than their subordinates) to be defensive (leaders' fragility), to set the ground rules (tone policing), to avoid an issue or make a joke out of it (silencing), or to disregard a trend, and suggest that the concern describes a rare circumstance (exceptionalism). These reactions to concerns – if not opposed – allow the leader to suggest that he is taking concerns seriously, whereas in reality, they are undermined or ignored.
8. … A traditional 'Just Culture'…: In traditional 'Just Culture', the measure of retribution is decided on the basis of a flow chart that requires someone to determine how culpable a 'perpetrator' is. See, for example, http://safetyandjustice.eu/ and https://deptmedicine.arizona.edu/patient-care/blog/quality-safety-%E2%80%98just-culture%E2%80%99-provides-process-review-correct-mistakes-optimal.

   For a further discussion on the drawbacks of a traditional 'Just Culture', see Dekker (2016). Note: the third edition is quite different to earlier versions.

   See for a further analysis of the determination of culpability, Cromie and Bott (2016).

   See also Heraghty, Rae and Dekker (2020).
9. … called 'Restorative Practice'…: I have chosen the term 'Restorative Practice', although this practice is often called 'Restorative Just Culture'. We have found the latter term confusing due to its similarity to 'Just Culture', while Restorative Practice is actually a very different approach. Dekker (2016).

10. ... According to some researchers...: Cook and Nemeth (2010).
11. Flinterman, M. (2019). Private communication, January 15th, 2019.
12. ... decisions are made...: In the fifties of the previous century, Herbert Simon (1955) challenged the contemporary models that assumed full rationality in economic choice. He suggested that human decision-making is limited by the incomplete information available for any practical decision and the prohibitive effort required to reduce uncertainty. As a leader, we need to acknowledge that our people operate with 'bounded rationality', not aiming for an optimal result but rather striving for a satisfactory outcome (called 'satisficing behaviour').

In the seventies and eighties, Tversky and Kahneman (1973) furthered this thinking by identifying a great number of biases that shape human decision-making. They explain how humans use mental shortcuts (called 'heuristics') to make decisions or form judgements, ignoring much of the larger or detailed picture.

According to Johnson-Laird (1983, 2006), humans can reason and solve problems because they construct simplified representations of the world around them in working memory. These mental models allow humans to operate effectively in routine situations by quickly grasping essential impressions with limited effort and saving cognitive resources for more complex tasks.

This thinking has been integrated into a new model of our brain. It is suggested that we have an autonomous, subconscious process (aptly called 'System 1') for routine tasks and a second process (System 2) for conscious considerations. System 1 thinking is characterised by Kahneman (2011) as automatic, implicit and unconscious. It requires little effort or brain capacity and is quite fast. System 2 is for controlled, explicit and conscious thought. It requires a lot of effort, is limited by our cognitive capacity, is slow and tends to be logical. System 1 is the default and switching to the more effortful System 2 is delayed as long as possible, until the cues from reality become so overbearing that we really can't delay any longer. System 1 is supposedly from primal origin, whereas System 2 is the result of more recent evolution and sets humans aside from other animals.

For a school of thought called 'Naturalistic Decision Making' (i.e. decision-making in a natural context instead of a laboratory environment) Klein, Orasanu, Calderwood and Zsambok (1993).

It seems that despite all our brain power, above all we are creatures of habit. Dennet (2013) supports such an assertion and suggests that our free will is not so much visible in any ultimate decision but much more in the time and effort invested to generate alternative courses of action. We might feel responsible for the decisions we have taken, but in reality, many actions come about through heuristics and experience. These theories help explain the results of the experiments by de Boer (2012) and de Boer, Heems and Hurts (2014) on the longer-than-expected time it takes for participants to change their problem-solving tactics or see a malfunction.

13. ... They only came to know...: Yves Morieux (2018, p. 6) puts this nicely:

"Instead of focusing on formal procedures, managers must pay attention to behavioral dynamics that shape organisational performance: why people do what they do; how they understand their individual goals, the resources available to them to realize those goals, the constraints that stand in their way; and how individual behaviors combine (often in unanticipated ways) to produce the collective behaviour underlying performance. What's more, because managers themselves are actors in the behavioural system, they need to know how to intervene in that system to foster more effective cooperation. And to that, they must get much closer to the actual work."

See also: Christensen, C. (2012). How will you measure your life? TEDx Boston, July 2012. Available at: https://www.youtube.com/watch?v=tvos4nO Rf_Y&feature=youtu.be, accessed March 2nd, 2020.

14. ... to engage in *sensemaking*...: Sensemaking is similar to analysis in that it is focused on taking decisions that are pertinent to a specific situation and is grounded in concrete reality. But it differs from ordinary analysis in that sensemaking happens on the fly without predefined procedures. It is often executed collaboratively with different interpretations of the situation appearing in quick succession. Sensemaking focuses on plausibility rather than accuracy and facilitates retrospection. It is these differences that make organisational sensemaking more powerful in a complex environment than analysis. Rankin, Woltjer and Field (2016); Weick, Sutcliffe and Obstfeld (2005); Kurtz (2014).

15. ... a detrimental effect...: Morieux and Tollman (2014, p. 81).

... Back in the 20s of the previous century...: The moral that was originally posited with this analysis was that if we 'prevent the accidents [then] the injuries will take care of themselves'. The author (Heinrich, 1941 in Busch 2019, p. 23) intended it as follows:

- Paying attention to 'small things' increases your knowledge base (frequency may be more important than severity).
- Potential is more important than actual consequence (a risk-based approach).
- You may prevent serious consequences from happening by reacting on events and conditions with no or minor consequences (a more proactive approach).
- For the purpose of accident prevention, all events are regarded equally important, regardless of their consequences.

See, for an example, of the current-day incorrect interpretation: https://www. skybrary.aero/index.php/Heinrich_Pyramid.

... Famous is the photo...: *New York Times*. (2014). Four Workers Are Killed in Gas Leak at Texas Chemical Plant, November 15th, 2014. https://www.nytimes. com/2014/11/16/us/four-workers-are-killed-in-gas-leak-at-texas-chemical-plant.html, accessed August 18th, 2019.

... Deepwater Horizon...: Examples like the Deepwater Horizon accident show that a focus on trivial safety actually distracts from serious issues. See Dekker (2019, pp. 121–131); Elkind Whitford and Burke (2011); Hopkins (2012).

16. ... Several recent studies...: Accidents that should be reported or counted, but that are swept under the carpet in an effort to retain an unblemished safety record are sometimes called 'Tipp-ex accidents'. See https://tippexongeval.be/; Barnett and Wang (2000); Saloniemi and Oksanen (1998); Storkersen, Antonsen and Kongsvik (2016); Dekker, S. (2018). Oil and gas safety in a post-truth world. Blog on https://safetydifferently.com/oil-and-gas-safety-in-a-post-truth-world/, dated May 18th, 2018, accessed February 26th, 2020.

17. Oil & gas industry, own experience, 2018. Following our debrief about our visit, efforts have been made to eliminate injury KPI's from the supplier management system.
18. … pursued zero harm targets…: Sherratt and Dainty (2017); Thomas (2020).
19. … Many companies struggle…: Unfortunately, organisations that provide a safe workplace have significantly *lower* probability of economic survival than companies that do not provide this. Safety does not come for free. Pagell, Parkinson, Veltri, Gray, Wiengarten, Louis and Fynes (2020).
20. … the confusing term 'safety culture'…: To suggest that 'safety culture' is a factor in improving psychological safety is a tautology. Indicators of both 'safety culture' and psychological safety are that staff (including novices) can speak up and discuss concerns openly. See for an example of such a tautology: O'donovan and Mcauliffe (2020). See also Dekker (2019, pp. 363–387); Guldenmund (2000, p. 11, 37); Henriqson, Schuler, van Winsen and Dekker (2014).
21. … those that were directly involved…: I promised not to use jargon so have avoided even the generally understandable synonym 'sharp end', which indicates where people are in direct contact with the safety-critical process, and therefore closest in time and geography to the (potential) adverse event. The opposite is the blunt end: the organisation or set of organisations that both support and constrain activities at the sharp end, isolated in time and location from the event. For instance, the planning and management actions that are related to a task. Dekker (2014b, p. 39).
    … a Qantas Airbus A380…: De Crespigny (2012).
22. … very limited predictive value…: Although some non-scientific sources suggest 'safety culture has been shown to be a key predictor of safety performance' (https://www.pslhub.org/learn/culture/good-practice/nes-safety-culture-discussion-cards-r2506/, accessed July 7th, 2020), science comes no further than a correlation (and often a tautology). Some suggest that a discussion on Safety Culture can enhance safety (https://www.skybrary.aero/index.php/File:Safety_culture_discussion_cards_-_Edition_2_-_20191218_-_hi_res_(without_print_marks)_Page_02.png, accessed July 7th, 2020), but this seems just as roundabout as talking about profitability to improve company profits. It can help create understanding but doesn't carry much weight in actual performance improvement.
23. … a nasty habit…: Quote from Hegde, S. (2020). Learning from everyday work. Blog for the Resilience Engineering Association, April 5th, 2020. https://www.resilience-engineering-association.org/blog/2020/04/05/learning-from-every-day-work/, accessed April 17th, 2020.

3

Alignment between Rules and Reality

3.1 Introduction

In this chapter, I will start to describe our proposed approach to safety by focusing on normal work rather than adverse outcomes. We do not rely on incidents as the trigger for learning. I will discuss the difference between how work is actually executed and how it is prescribed in rules and procedures. We assume that people use rules as a resource to fulfil their goals and may interpret them differently than you intended. There are good reasons why there might be a lack of compliance in your organisation – quite possibly even on a daily basis. The easiest way to get alignment between the rules and how work is done in reality is to get the workforce to draw up the procedures, and help you maintain them.[1]

There is often a difference between the rules and reality, but not because of malicious intent. Actual work is done under the duress of conflicting goals, and sometimes with a lack of resources and tools, and occasionally with gaps in training, knowledge and experience. Procedures and rules are drawn up without the intricate knowledge of everyday work and with assumed close to ideal circumstances. In the actual execution of tasks, this variability is taken in our stride and compensated for so that nearly always a successful outcome is achieved.

> In March 2017, an Airbus A340 aircraft taking off from the high-altitude airport at Bogota, Colombia only just managed to become airborne before the end of the runway and narrowly missed hitting antennas at the runway edge. The investigation showed that the pilots had followed the instructions correctly, yet the result was that the aircraft was too slow in lifting off. Looking back in time, the investigators found out that this was a common problem with this aircraft type across multiple airlines at this airport which had not been previously identified due to a lack of reports. Although there had (as yet) not been an accident, this was not a situation that could be endured. Therefore, operational instructions were amended, and targeted training was given to avoid this dangerous situation in future.[2]

This example shows that some rules – even in aviation – are not appropriate, but that mostly successful outcomes are achieved regardless. We are not always so lucky. Five crew members perished in a fire on board the

Maersk Honam, one of Maersk's ultra-large containerships, in the Arabian Sea in 2018 while en route from Singapore to Egypt. After the incident, Maersk Line announced that a thorough review of the then-current safety practices and policies in the stowage of dangerous cargo identified that a new guidelines were required to improve safety across its container vessel fleet. Apparently, in this case, this need had not become so obvious earlier.[3]

These examples challenge traditional safety thinking that rules are perfect and should be abided by, and that any infringement constitutes an error or – worse, in case of intent – a violation. If we assume that people are not pathological, then there must be some logic that triggers the infringement (whether accidental or intentional). This might include production pressures (e.g. the turn-around time of an aircraft at the gate), personal gain (e.g. getting home on time, less effort), situational factors (e.g. poor design of tools or incorrect procedures), lack of knowledge (e.g. as a result of poor training), personal factors (e.g. fatigue), cognitive factors (e.g. distraction) or task-related factors (e.g. poor interface design). Note that operators will always need to make a trade-off between many different objectives, of which safety is just one – if that.

In the rest of this chapter, I will first discuss the rules (in the safety literature called 'Work-as-Imagined', or also *paper* in this book) and reality (called 'Work-as-Done', or *practice*) to understand the value of each. I then discuss how to identify differences between the two and understand why this gap exists. Only then are we able to close the gap and – importantly – ensure it remains small by making your people feel accountable for keeping it closed.

3.2 Work-as-Imagined

Work-as-Imagined is what designers, managers, regulators and authorities believe happens or should happen. Work-as-Imagined is the basis for design, training and control. Work-as-Imagined is captured in process diagrams, work instructions, procedures, ad hoc directives and verbal instructions. Task formalisation – although 'imagined' – is important. Many tasks consist of multiple steps, and descriptions are required to help people remember, particularly where time is limited. Task descriptions help us to educate and train people for their jobs. Defined tasks are necessary to ensure that people can cooperate effectively. Task descriptions are necessary for design and planning purposes within an organisation, and as a means to identify variances in behaviour that lead to unacceptable risks.[4]

However, we often see that upon close scrutiny, task descriptions likely contain contradictions and superfluous details. Taken together, all the documentation for a given task is often much more than is reasonable. An inflated reliance on Work-as-Imagined is based on the outdated Western paradigm that complete knowledge is possible, and harm is foreseeable. As labour unions well know, hardly any work gets done at all if we strictly follow procedures: in dictionaries, 'work-to-rule' is actually classified as an industrial action – also called 'malicious compliance' for good reason. So before we go about making everybody conform to impossible procedures, let's really understand what our people are up against in practice.[5]

> If an aircraft is not stabilised according to well-defined criteria in its approach at 1000 feet above the runway, the Standard Operating Procedures of most airlines specify that the landing is to be aborted and a go-around executed. One report suggests that over half of all aircraft accidents can be avoided if these rules are adhered to. But in fact, in only 3% of the non-stabilised approaches a go-around is actually executed. In the vast majority of cases the crew persists in the approach and landing, and only in a very few isolated cases an accident occurs. The non-compliance in some ways actually aids safety, because a go-around and a second approach introduces new hazards to the aircraft and the wider aviation system. In going around, the aircraft may again be unstabilised at 1000 feet, necessitating yet another go-around while fuel is depleting and passengers are delayed. In a study at one airline, we found that there were two main reasons for not complying with the instructions. At some airports in mountainous terrain it is nearly impossible to comply with the stability criteria due to the curved flight path towards the airport and gusty winds. In many other cases it emerged that pilots have their own, less stringent definition of a genuinely non-stable approach which is much more difficult to capture in simple criteria. This definition turned out to be consistent amongst pilots and instructors and was implicitly applied by all of them at this airline in the pursuit of fuel efficiency and timeliness.[6]

This example illustrates the impracticality to adhere to seemingly simple rules like this all the time under production pressures and environmental circumstances.

Work-as-Imagined is dislocated in time and place from execution and is therefore lacking immediate feedback. The circumstances surrounding Work-as-Imagined are assumed and static and will not always reflect the everyday variability and sometimes extreme circumstances under which work is done. The resources for Work-as-Imagined are predetermined and of assumed predictability. Training, knowledge, attention and motivation are all assumed to be appropriate, whereas during actual task execution, this might not be the case.

Up to a hundred years ago, task descriptions were very uncommon. Artefacts were made by craftsmen who were autonomously responsible for the production process, tooling and the quality of the end product. With the advent of Taylorism in the 1930s, jobs were partitioned and a description of every task was created for design and planning purposes, to train people and as a means to identify variances in behaviour. These descriptions were created by a bureaucracy that allowed no input from the workers themselves and had absolute confidence in their own ability to imagine the requirements of the task at hand. These days, organisations are comprised of professionals that have a thorough understanding of what is expected of them and how to do their jobs. The difference with the era of craftsmen is that we cannot do our jobs alone. We are always part of an (extended) organisation, and our own processes are intertwined with those of others. In all but a few isolated cases, there is a strong need to cooperate with those to achieve common goals. We are collectively responsible for the quality of our processes and the outputs that are generated, and our organisations and its managers – not the individual professionals – are generally held accountable for employee well-being and public safety. This requires us to establish and uphold descriptions of how work is to be done, identify the gaps that occur between *paper* and *practice*, and decide what – if any – actions are warranted because of this gap. Individual competences are still important, but we now label these 'professional skills' rather than 'craftsmanship'.[7]

3.3 Work-as-Done

Work-as-Done is what people have to do to get the job done in the actual situation. Work-as-Done is what actually happens. Availability of resources (time, manpower, materials, information, etc.) may be limited and uncertain. People adjust what they do to match the situation, but these adjustments will always be approximate. Remember that our behaviour is triggered much more through heuristics and experience rather than conscious thought, even though we feel responsible for our actions. Performance variability is inevitable, omnipresent and necessary. Performance variability is the reason why things sometimes go wrong, but it is also the reason why everyday work is safe and effective. Work-as-Done is 'in the moment', that is to say it is being executed at a specific place and time and isolated from reflection before or after the fact. Each time a task is performed, it will deviate from its previous execution. There are as many versions of Work-as-Done as there are instances of the task, and so there will be a multitude of justifications for the gap between *paper* and *practice*.[8]

> At a major European airport, we investigated the compliance to the rules for ground services: the activities to turn-around an aircraft after arrival, including disembarking, unloading, cleaning, fuelling, loading and boarding. Contrary to what one might expect in aviation, we found a large percentage of non-compliances such as: equipment not parked in regulatory areas (68%), wheel chocks not correctly placed (44%) and hearing protection not worn (37%). In all observed cases (i.e. 100%) the safety barriers alongside the conveyer belts leading up to the hold were not used. This led to baggage falling on employees and lost-time accidents in four specific cases. Other common errors included not checking

the hold after unloading (69%), setting conveyor belt height during driving (62%), and employees walking on a running conveyer belt (48%). In 40% of the observations the chocks were removed before connecting the pushback vehicle. A pre-departure check was not performed correctly in 29% of the observations.[9]

These examples show that even in the highly regulated aviation industry departures from the rules are common. Furthermore, given the high frequency of these deviations, they can be deemed systematic rather than personal. There must be a logic to these nonconformances that we can identify and use to close the gap. However, it was difficult for management at the ground services company to make any progress because they approached the breaches from a normative standpoint ('the procedures are correct and need to be adhered to'). They were initially not able to ask open questions or to convey authentic curiosity about Work-as-Done in their organisation. Management at this company (as at many others) is prone to intense pressures on productivity, timeliness, cost containment and 'risk management' – the latter relating to compliance and risk administration more than addressing unknown hazards – which all constrain the willingness to know about the dirty details of actual work.[10]

Contrary to common belief, Work-as-Imagined is not even always safer than Work-as-Done:

> In a study of European shunting yards, we found multiple examples in which in actual work was safer than on paper. For example, surveillance cameras to oversee a shunting yard were available in practice but were not described in the procedures. These helped the shunting yard managers to identify where the railway equipment was in the yard and to determine whether the tracks were clear. These were used to supplement the procedures in ensuring safe operations.[11]

Although Work-as-Done is sometimes safer than Work-as-Imagined, in both a review of the literature and through a survey among aviation companies we found that the *size* of the gap between Work-as-Done and Work-as-Imagined is considered relevant to safety. If the gap between procedures and practice is small, we can be sure that all the thinking that has gone into devising the procedures is being used. A large gap between rules and what happens in practice is undesirable, because some of the assurances that procedures offer are systematically undermined. After all, rules, regulations, procedures and such have a purpose. If the gap is large, risks might be introduced that we had considered mitigated, for instance, related to high-impact low-frequency events such as explosions or aircraft crashes. These may not have been experienced by the workforce themselves due to their sporadic nature, and therefore are not part of their collective memory. So naturally, we need to take exceptions from the written rules seriously.[12]

It seems beneficial for safety (and probably performance) to close the gap between *practice* and *paper*, but not necessarily in the direction of

Work-as-Imagined. The gap cannot be closed simply by enforcing the procedures – after all, why are they not being abided by in the first place? Apparently, there is something stopping our people from adhering to procedures, and we first need to understand what that is. By emphasising the need to comply with rules and regulations, we are preserving the very factors that lead to a gap and are driving deviations from the rules underground. To enable a frank and open conversation about Work-as-Done, it is necessary to emit a sense of genuine curiosity that excludes any normative attitudes. We perhaps actually need to appreciate (!) the gap between *paper* and *practice* because that is what allows us to learn and improve. But how do we get to know about a gap in the first place?

3.4 The Elusive Gap

Seeing the gap between rules and reality can actually sometimes be surprisingly difficult.

> The work instructions for a filter change on an oil & gas platform in the North Sea consisted of 23 pages of text, and about 100 separate instructions, each of which had to be signed separately upon completion of each individual task. The filters were changed and cleaned multiple times each week by experienced control room operators, who needed to go out in the cold and the wet. We didn't believe that they took the instructions with them to sign after each step, especially in bad weather, and after a bit of challenge it turned out that the team usually signed off the whole workbook of instructions back in the control room after the task had been completed – which made the sign off per line item pretty pointless. Getting the control room operators to own up about this was not easy, as they weren't about to jeopardise their good standing with management about it. And why would they – management was happy that there was a tick in every box, and the operators got to keep their gloves on while they were outside. Who needs to rock the boat for something so trivial? It was only after talking informally with them for some time, in the night shift, that we finally heard this admission.[13]

Generally, you will find that operators make adaptations to the way that work is done to absorb an increase in workload and achieve a multitude of objectives. These adaptations are not really visible in superficial observations but require a deeper scrutiny and real conversations.[14]

Often incidents are an indication of a systematic gap between *paper* and *practice*. But even after an accident occurs, it takes a bit of perseverance to identify that certain ways of working have gone unnoticed for quite a while.

> In one instance management was completely baffled that someone would use a rubber hose to vent a high-pressure process installation - leading

to uncontrolled whipping of the hose and injury to the technician. The incident investigator went to great lengths to explain to us that it was against all regulations and even counter to common sense to use a flexible hose for such a task. But after the investigator had left the meeting, the team leader conceded that rubber hoses were actually used quite regularly to vent process equipment in an effort to save time and ensure that any residual liquids were properly caught.[15]

A word of caution: incidents by definition imply an unwanted outcome, often surrounded with emotions, guilt and blame, and so there is a tendency to judge the preceding actions in hindsight. That makes it much more difficult to extract learnings from them than routine work, but still the opportunity should not go to waste. Actually, for the purposes of identifying a gap between Work-as-Done and Work-as-Imagined, we are not interested in the incident itself, but instances just like it that did *not* lead to an adverse outcome. Again and again, investigations have identified that before a major catastrophe, similar situations had arisen earlier from which learnings might have been distilled. Incidents are an opportunity for us to identify situations where work was clearly challenging and investigate instances just like it that have led to success. As the example about the take-off in Bogota at the beginning of this chapter showed, there is much value in assessing a near-miss, and – especially – incorporating previous routine events in that investigation. It helps us to identify where work is challenging and where safety is sacrificed before adverse events occur. A culture of 'accountability' and focus on the individuals who were directly involved (rather than planning, management and other support processes) are not conducive to seeing the gap. In fact, a consistent gap between *paper* and *practice* should set us thinking not just about the circumstances surrounding a particular incident, but particularly about the lack of awareness with management about what really happens on the shop floor.[16]

You can't always rely on audits and inspections to identify a gap between *paper* and *practice*, because they only cover what you think needs to be covered:

> In January 2020, the Catharina Hospital in Eindhoven reached out to 649 former patients to have their blood tested for hepatitis B, hepatitis C and HIV which they could have contracted at the hospital. The hospital used a specific instrument called an ultrasonic-tip to remove dental tartar, a treatment that sometimes causes the gums to bleed slightly. Thermal sterilisation is the appropriate method to disinfect the instrument, but in contradiction to the hygiene guidelines the equipment had only been sanitised with alcohol since January 2014. The hospital could not exclude that contaminated blood from one patient was transferred to others, possibly infecting them with a virus. In a statement, the hospital says that it is organised to maintain hospital protocols but that dental cleanings are covered by 'dentistry' and are executed "out of sight of all hospital checks". After it became known that the hygiene guidelines had not been followed, all treatment with the ultrasonic tip was discontinued.[17]

In this case, the dentistry activities were not initially considered to be subject to hospital protocols and so the gap between actual work and the appropriate hygiene standards was never identified.

You can't even always rely on reporting systems to relay important safety issues, either:

> In 2007 a McDonnell Douglas MD-83 deviated from the approach path to Dublin Airport because they mistakenly took the lights of a newly constructed hotel for those of the runway. Upon coming closer to the hotel (and other obstructions on the ground!) they wondered about the strange lighting pattern and asked air traffic control about this – still under the impression that they were aiming for the runway. Luckily, from the radar the controller realised their deviation and was able to make the crew abort the approach. After a go-around and a safe landing, the pilots were not inclined to report the incident. The air traffic controller reported it as a non-serious incident. Only after publications in the media about the low flying aircraft did the Irish Air Accident Investigation Unit initiate an enquiry, which finally led to changes in the lighting infrastructure of the hotel so that in future confusion is avoided.[18]

Seeing the gap is even more difficult when multiple parties are involved, and when these parties trust the other to know what is really going on.

> On Saturday August 10th, 2019, while a storm is raging, part of the roof of the football stadium in Alkmaar, the Netherlands collapses. Luckily no one is injured as the stadium was unoccupied at the time. Inspections in the days after the accident indicate that the steel construction collapsed due to poor welding. The stadium is only 13 years old and the structural integrity had been approved each year. It quickly turns out however, that the annual approval is based entirely on a paper exercise in which the Dutch football association, the city council, the fire brigade and even the home office are holding each other accountable for the safety of the spectators.[19]

As each of these examples show, it is actually quite easy to overlook gaps between *paper* and *practice*. Work is already tedious enough not to be bothered about little things like that... Still, we need to know about gaps. There are a number of ways to get a sense of where gaps between *paper* and *practice* exist – in fact, once you get a nose for it, you will find these gaps much more easily.

3.5 Identifying Gaps

A good way to appreciate the gap between Work-as-Done and Work-as-Imagined is to go and have a look for yourself. Most managers' days are jam-packed with meetings, and the remaining time is filled with reading and preparing for those. Nevertheless, the age-old adage about the merits of management-by-walking-around still hold. You'll get some feel for how people are handling the variability of the real world, and you might even get a chance to talk to your colleagues about how they are coping with the balance between efficiency and thoroughness. That's not the way to see the whole picture though. You'll rarely see someone consciously breaking the rules when senior staff is around, and you will be lucky to see those exceptions where a work-around is needed on the spot. We have found that meeting the night shift for a few hours can be very insightful – having a jet lag helps! An alternative to walking around yourself is to have the operational experts come to you and explain how work is done. This allows for a reflective atmosphere but limits the physical experience of the job. As external consultants/academics, we have often been able to open up the conversation with front-line workers much more than any senior manager would be able to. We were able to spend a good deal of time with them and were clear about our intentions to understand real work, and not evaluate individuals. We were honest about the things we noticed and shared these with them as a matter of course. One particular novel way to identify the gap between Work-as-Done and Work-as-Imagined is to ask for narratives: short stories that have been experienced by the storyteller himself/herself and that can be collected through an app on a smart phone or tablet.[20]

So how can you open up the conversation and find out about gaps? Well, you might be interested to know:[21]

- When is this task difficult?
- What are you dependent on to do a good job?
- Tell me about a good day.
- Are tools and resources always available to do the job?
- What do you do if you can't access tools?
- And resources in time?
- What solutions have you come up with that the rest of the organisation could learn or help you to improve?
- Where are we wasting time/money?
- Is there something which is nonsensical or unnecessary that you have to do here?
- What is the first thing you sacrifice when the going gets tough?

Note that not once reference is made to the procedures or work instructions. If these are helpful for the task, they will be mentioned. And if the operator doesn't mention them, well, then they just might not be that useful... One of our aims is to have a shared understanding of the reasons for a possible deviation between *paper* and *practice*, and under which circumstances this is deemed necessary to achieve the desired objectives – and what these objectives might be. Often, the objectives that were implicit (imagined!) at the time of the task design are no longer completely valid, for instance, due to a modified emphasis on productivity goals. Your people need to break the rules because the world is more complex and dynamic than the rules. The way for them to get work done is to keep learning and adapting effectively and safely to varying circumstances.

Sometimes there will be areas in your organisation where you will be uneasy about how things get done. Don't turn a blind eye but see that as an opportunity to discover more. It needn't be the operational front line exclusively; you could equally focus on engineering or maintenance. It helps if you can find people that are equally inquisitive and do not react defensively when confronted with a gap between Work-as-Done and Work-as-Imagined. Try to get a feel for where the real risks reside, and don't be misled into identifying petty gaps that only serve to distract.

One of the questions that we regularly get asked is: 'Where do I get the biggest bang for the buck? Where do I start?' Our simple answer is: Where would you be most embarrassed to find a significant gap between *practice* and *paper*? That seems like the logical place to start. The next question we often get is: 'Oh boy, there are so many gaps, how do I set priorities?' Well, that's an easier one: don't set priorities but get all of them addressed. To do this, you need to ensure that the gaps are not being addressed by yourself or safety personnel, but by line management. This will take a bit of coaxing and coaching, because everybody except you will want the safety department to act on this. Then line managers don't have to take actions or assume responsibility, and the safety staff can show how important they are. But you'll get half-hearted results because the gap needs to be owned by the operational team, not some staff department, and progress will be too slow. You can quite easily make the line managers accountable for any gaps that you hear about and that are not properly managed, as is shown in the next section. A sure way to get started is to expect your direct reports to tell you about gaps between *paper* and *practice* before you hear about it from others. That way, you get to leverage the potency of the whole organisation, not just that of a few individuals.[22]

In this section, I discussed ways to identify gaps between *paper* and *practice* such as voluntary reports and incident investigations. Consider again the case of the road worker introduced in the previous chapter. There is no way that the planning errors necessitating the crossing of the motorway would have become apparent if the inspectors had not been really curious why it made sense for the road worker to cross the road and instead had started the conversation by scolding him for his 'carelessness'. It is possible to have this discussion in an ad hoc fashion while walking around, but it can also be very effective to include this type of questioning in the lowest level (self-)audits, which should be focused on maximising learning rather than minimising findings. To identify gaps, we need to be curious about how work is actually done and to have an appetite for bad news, even if it turns out to be embarrassing. Maybe there is consolation in the thought that the shame is always less now than it will be after an adverse event.[23]

And then, when you know about a gap, you will need to do something about it – after all, legally and morally, you cannot leave a significant breach of procedures go unheeded.

3.6 Closing the Gap

To start closing the gap, it's good to get a few people into a room and discuss the examples that you have witnessed or heard about. The end goal is a shared and detailed picture of the task under different circumstances, and agreement about the documentation (rules, guidelines, checklists, etc.) that will best support the execution in practice. It seems only logical to involve the people that are affected by the rules in their design, as well as domain experts. [24]

> In the example of the work instructions for the filter change, we held a workshop with team members and central engineering staff. It took a while for the front-line operators to have a voice, initially everyone was happy for the staff members to do all the talking. But after some coaxing the operators chipped in, and we were able to define the critical check points that we (collectively) considered to be so important that we wanted them signed off. We also considered which guidance material might be useful for an operator that was doing the change for the first time, taking into account that they would always be accompanied by a more experienced colleague. The end result of this workshop was a procedure that was agreed by all and less than half in length than the original. It was implemented successfully and in fact a few months later the regulator got to hear about the new procedure and indicated their support for the approach and the result.[25]

Our approach assumes that there is enough trust and respect between parties to enable this discussion. The safety department can be called upon to not just add their expertise, but also facilitate the discussion, and help to strike the right equilibrium between the required detailing of the rules and the autonomy to adapt to varying circumstances. If staff and management are up to it – that is to say they will not judge motives – they can be included to learn and understand. However, you may find that they will quickly fall into the trap of defending the current rules, procedures, checklists, standards, job descriptions and management systems, rather than taking Work-as-Done as the reference and seeing where Work-as-Imagined falls short of helping to achieve a particular purpose in a particular context. So in that case, you will need to remind them of this...[26]

When we have a common understanding of the reasons that *practice* deviates from *paper*, we can make a start with closing the gap. We strive for agreement about the extent to which task formalisation is required versus autonomy for the operators to cope with process and external variances. We want the gap between *paper* and *practice* to be balanced between organisational needs and the variances encountered in practice – and that this gap is defendable against external scrutiny such as audits. This balance may vary between locations or teams: one size does not necessarily fit all. On the one hand, a set of procedures is needed to help people remember the steps under challenging circumstances, to educate and train people for their jobs, to ensure that people can cooperate effectively, for design and planning purposes, and as a means to identify variances in behaviour. On the other hand, we should not excessively restrict people to cope with the variances that they routinely encounter in practice; that is, exemptions to the rules should be justified and limited in number. The appropriate level of detail will vary from context to context and is dependent on several factors. More detail in the rules will support the organisational needs (help people remember the steps and cooperate, training of operators, for design and planning purposes, and for the identification of variances) but will increase the number of exceptions to be managed, particularly if these tasks occur in strongly differing conditions. As an organisation matures and the operators become more competent, we migrate to framework rules within which the professional has freedom to execute his tasks (we sometimes call this *freedom in a frame*). If the workforce is expanded or rejuvenated, we may need to revert back to more detailed work descriptions. Note that there may be other solutions for aligning Work-as-Done with procedures than just changing the procedures. We may be able to add physical barriers, redesign the workplace or change the way that tasks are allocated across a group of people. In fact, these redesigns are generally preferable to the limited efficacy of written instructions to support people in executing a task.[27]

A useful process for aligning *paper* with *practice* is given in Figure 3.1. (1) We start by having a really good understanding how work is done in practice. Feedback from the front line allows us to comprehend where the rules and procedures are working, and where they need to be improved. (2) As suggested before, the monitoring of the use of the rules should not evolve from a requirement for compliance but a real need to learn. In evaluating whether a rule is currently satisfactory or needs to be changed, we need to understand

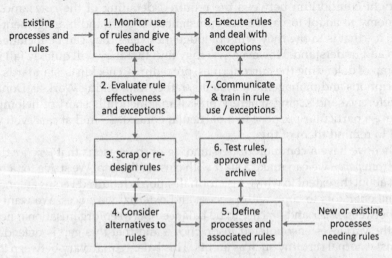

FIGURE 3.1
Flow chart for rules management. (Adapted from Hale and Borys (2013b).)

how work is done in practice and how the rules support the front line in varying conditions. (3) The rules might be too detailed for the variance that occurs in practice and the maturity of the workforce (like the extensive step-by-step filter instructions were not compatible with North Sea weather), so that (parts of) rules are superfluous and need to be scrapped. This also allows us to redesign and reword rules so that they are more suitable for the tasks they are intended to support. (4) There may be other solutions for aligning Work-as-Done with procedures than just changing the procedures that need to be considered. (5) The need for new rules requires us to draw them up for the first time, possibly considering alternatives to rules as well. (6) The implementation of rules requires us to first test them (see, for instance, Chapter 4), then approve them and ensure that they are stored and made available through some sort of organisational memory system. (7) Appropriate implementation of rules is necessary to ensure that those that we expect to apply the rules know about them – and ideally buy into them as well. We need to communicate the changes to the rules and train people in their use – including how we expect them to manage exceptions when they feel they cannot abide by them. (8) Only when we have communicated the new rules and achieved buy-in can we expect our employees to apply the rules and deal satisfactorily with exceptions.[28]

Some good guidelines for drafting the procedures are as follows:

- Use 'swim lanes' (one per accountable role) to depict the process and its associated rules in a flow chart. Try to limit each process to a flow chart on a single page. Details of the process steps and constraints can be added as text on additional pages. If the flow chart doesn't fit on a single page, split the process into multiple flows.
- Ensure that the enforcers of the procedures (those that determine the compliance) are not the makers. Definitely include their input but ensure that they do not feel so much ownership for the procedures that they cannot kill their darlings or consider every divergence from the rules a violation rather than an exception.
- Consider that every task (and therefore its description) needs to satisfy multiple goals. Integrate these goals into the process description, and ensure that a task is not subject to multiple, possibly contradictory descriptions. Don't let your workforce identify and solve inconsistencies on the spot.
- When considering (an update to) a procedure, ensure that obsolete rules are eliminated. Technology or policy changes can make the existing rules irrelevant – or worse, counterproductive. Rule creep may have occurred if rules have been given a broader application than their original intent.
- Include the reporting and monitoring requirements in the procedure, and ensure a balance between organisational needs and the disruption that is caused by reporting. Eliminate duplication.

- Contemplate alternatives to rules and procedures to improve performance and reduce risks, such as specific tooling, redesign of processes and physical barriers. Also consider the use of 'nudges' to ensure more compliance with good practice, unwritten cultural rules or written instructions. For example, to reduce shipping damages, an up-market bicycle manufacturer printed a television on their boxes.
- Pilot the new rules before giving the final go and keep monitoring their use even after this.
- If you find that you are continuously coping with exceptions, then the procedures are too detailed. If you find that the organisational needs (memory aid, training, collaboration, design and planning and monitoring behaviour) are not met, then further detailing is warranted. Getting the balance right is the challenge.
- Some organisations put a 'next review' date on procedures to ensure that they are revisited regularly. We find that this rarely achieves its aim but rather burdens valuable resources needlessly. Some rules will require frequent reconsideration even before the review date, whereas others can remain unchanged for much longer. Other triggers such too many exemptions or not meeting organisational needs are more effective in identifying the requirement for improved rules.

Of course, in drafting the procedures, the people that actually execute the work are in the lead, as they understand Work-as-Done. Staff members such as safety professionals contribute by helping to identify and analyse risks and suggesting measures to mitigate these. In most companies, operational management has the final say regarding the procedures, taking into account the proposal from the shop floor and the recommendations of the safety staff. Hopefully, there is a shared understanding that the final version of the rules (*paper*) balances risks with workability and variation with traceability – and so best meets the various objectives and multiple instances of *practice*.

3.7 Maintaining Alignment between Rules and Reality

At this stage, we should have procedures that are as close to 'regular' instances of Work-as-Done as is reasonably possible, and that fulfil the organisational goals of Work-as-Imagined described before (memory aid, training, collaboration, design and planning and monitoring behaviour). These rules have been implemented, which really means to communicate the new procedures (and their justification), train the task executors if required and make the procedures readily accessible for when they are needed.

The challenge now is to make people feel accountable for signalling when they can't get their job done properly. Any significant deviation from the rules should be infrequent and feel uncomfortable, both for the task executors and for their supervisors, yet there is a shared understanding that sometimes exceptions are necessary. We should attempt to introduce an attitude where these exemptions are recognised and discussed – either before the fact if possible, or afterwards. We can then agree whether these exceptions:

- Are due to truly exceptional circumstances that do not warrant further action;
- Require an update in the procedures that we need to address;
- Are indicative of poor implementation (communication, training, access, buy-in).

In the end, it is our corporate responsibility to be aware of these exemptions, and to collectively agree in which category they fall. We should not – and increasingly cannot – hide behind the façade of not knowing.

What we would like to achieve is that the operators when they need to deviate from the procedures signal this to peers or supervisors. Front-line employees feel responsible for adherence to the procedures and deviate from them at their own discretion but knowingly and justly. They take responsibility for these exemptions and are willing and able to explain their considerations. There is effective peer pressure to abide by the rules unless the circumstances dictate otherwise. It stands to reason that this is only possible if workers accept the regulatory framework as a useful basis from which to assess what needs to be done in practice and also feel trusted to make the right choices when they are executing their tasks. Real-life examples can be used in workshops to uncover assumptions, values, beliefs and decision processes and to help grow the number of professional discussions on the gap between *paper* and *practice*. Repeated exceptions generally signal the need to reconsider the procedures, implying a return to earlier steps in the flow chart. Repeated exceptions that are not highlighted voluntarily within organisations are an indication of a lack of transparency and psychological safety, and a need for management to be more welcoming to bad news. A separate word is warranted about operators outside your own organisation. Long-term subcontractors can usually be regarded as own employees, although the subcontracting agreement can confound a collaborative attitude. We have heard of subcontractors that are scored on the number of incidents or 'violations' that occur – as you can imagine that doesn't create a willingness to report exceptions.[29]

In an environment of low social cohesion, it is more difficult to make (autonomous) operators abide by rules and feel accountable for exceptions to procedures. Examples include cab, truck and forklift drivers, independent professional pilots and small-group mariners – and of course consumer-operators like car drivers, recreational sailors and general aviation pilots.

These may be easily seduced to violate rules solely for personal motives. In these cases, we still need the rules to be properly implemented (communicate the new procedures and their justification, train the task executors if required and make the procedures readily accessible). Subsequently, the adherence to rules is supported by an enforcement system of minor retribution (e.g. fines) or rewards. For this to be successful, the system needs to create a high probability of detection, equal treatment of perpetrators independent of outcome, reasonably sized penalties for infringements and a reliable collection of levies. The penalty should be administered independent of the consequences and not exclusively if the infringement leads to an accident. Two examples of such systems are permanent automatic measures to counter speeding violations using radar and cameras over a trajectory, and a system to ensure garbage is placed in (and not next to) the street-side containers.[30]

In most organisations, there are formalised processes such as audits that are intended as a (secondary) means for confirming that Work-as-Imagined and Work-as-Done are closely aligned. Findings and other comments from auditors can be countered and discussed with self-confidence, and justified findings are considered triggers for learning rather than as failures – and hardly a cause for embarrassment. Ideally, the organisation (not the auditors!) decide and are able to justify how closely aligned *paper* and *practice* should be.[31]

Audits can only generate valuable information if the urge is suppressed to minimise findings and conceal gaps. Take this recent e-mail example that is possibly representative for many organisations:

> From: …
>   Sent: [date]
>   To: …
>   Subject: Fire Brigade Inspection
>   Importance: High
>   Dear all, [The local] Fire and Rescue Service will be carrying out an inspection of [building] today [date] at [time]. Can you please ensure that work areas, walkways, emergency exits, server rooms and stairs are clear from obstruction. Your support is appreciated.
>   Regards, …,[32]

This e-mail thwarts all efforts at safety because it is focused on minimising findings rather than using the external audit as a secondary measure in the support of the alignment of reality and rules and to confirm successful exception management. Business-as-usual should stand up to internal and external scrutiny. Audits rightly take written procedures as the standard and check compliance against that, in the justified aim to find out whether the gap between Work-as-Done and Work-as-Imagined is being managed. Auditors expect you to have at least achieved a certain level of safety leadership. Any findings help us learn about our safety practice and should not be concealed in an effort to avoid embarrassment.

3.8 The Soft Skills

For employees and managers alike, it will be a significant paradigm shift to consider rules of one's own creation and jurisdiction instead of belonging to an external agent. The change will require organisational embedding which can be facilitated by the safety department, and behavioural changes in the field. Although a generic guide to triggering and maintaining behavioural changes is outside the scope of this book, I offer some additional explanation in the specific context of eliminating gaps, taking ownership of the rules and feeling responsible for any remaining exemptions.[33]

Changes in behaviour are triggered by the realisation that this is unavoidable: even though it is extra work, there is no escaping the necessity to eliminate gaps and help draft new procedures. Taking ownership of the rules should be more attractive than having rules imposed by outsiders, for instance, because these procedures are easier to work with, because there is more autonomy in defining exemptions and because non-compliances are considered as a learning opportunity rather than a potential for punishment. The change needs to be facilitated by a physical infrastructure that minimises the effort that is required to accommodate the new responsibilities. Initially, it will be necessary to repetitively reinforce the new ways of working: not just in the field but also with those that have suffered the loss of direct responsibility for rules. The reinforcements will focus on the new *process* of rule creation and maintenance, and not (just) the exemptions. You need to keep asking how new rules came about, and how exemptions that you hear of have been categorised into truly exceptional circumstances, requiring a rule update, or poor implementation. You will be addressing whole departments and their managers rather than a specific operator, being curious rather than judgemental. You need to do this publicly to allow the message to sink in with the silent majority, so that eventually the new way of working is supported by social norms.[34]

3.9 Conclusion

In this chapter, I have addressed the gap between Work-as-Imagined and Work-as-Done. I have suggested ways to close the gap by first of all ensuring that the gap is identified, and then understanding why the gap makes sense to those who are out there doing the work. I have suggested a simple process to close the gap and have shown how to involve your team members to keep the gap small. The steps that I have presented in this chapter lead to rules and procedures that as closely as possible match the various instances of Work-as-Done. The procedures balance the need for local adaptation with the organisation's desire for a nominal way of doing things. They are generated by taking the operator's view as starting point without discounting any significant hazards or ignoring available expertise. Task executors feel accountable for maintaining alignment between Work-as-Done and Work-as-Imagined and to signal whenever exceptions are justified. We are able to justify these exceptions to internal and external observers.

Key actions:
- Appreciate the gap between work as it is actually executed, and rules and procedures as an opportunity to learn. Emit a sense of curiosity and avoid a normative attitude.
- Have conversations with the front line about how work gets done and what sacrifices are made.
- Consider asking externals (even students) to describe how work is really done.
- Expect your direct reports to tell you about gaps between rules and real work before you find out for yourself.
- Expect your direct reports to understand these gaps, close these gaps with the front line, and maintain alignment.
- Expect your reports to properly implement new or modified rules.
- Use audits as a secondary measure in the support of the alignment of real work and rules to confirm successful exception management.
- Ensure that rules strike the right balance between meeting organisational needs and minimising exceptions. Monitor this balance in the face of changes in workforce maturity and external circumstances.
- Repeatedly ask about the process of rulemaking and exception management to reinforce the approach.

Although in this chapter I have addressed the process of closing the gap between rules and reality, sometimes it is not immediately evident which measures will be most effective. We might need to try a few different things to see what works. A good and safe way to do this is to innovate and improve using micro-experiments: the subject of the next chapter.

Notes

1. How work is prescribed is referred to in contemporary safety literature as 'Work-as-Imagined' or in this book as 'rules' or 'paper' for shorthand. How work is actually executed is labelled 'Work-as-Done', and in this book as 'reality' or 'practice'.

2. Bureau d'Enquêtes et d'Analyses pour la sécurité de l'aviation civile (2019).
3. ... a fire on board the Maersk Honam...: Manaadiar (2018), Furness (2020).
4. ... Task formalisation [...] is important...: Hale and Borys (2013a).
5. ... 'work-to-rule'...: https://www.merriam-webster.com/dictionary/work-to-rule; https:// dictionary.cambridge.org/dictionary/english/work-to-rule.
6. The Flight Safety Foundation suggests that 'all flights must be stabilised by 1000 feet above airport elevation in IMC and 500 feet above airport elevation in VMC. An approach is stabilised when all of the following criteria are met:

   • The aircraft is on the correct flight path.
   • Only small changes in heading/pitch are necessary to maintain the correct flight path.
   • The airspeed is not more than VREF + 20kts indicated speed and not less than VREF.
   • The aircraft is in the correct landing configuration.
   • Sink rate is no greater than 1000 feet/minute; if an approach requires a sink rate greater than 1000 feet/minute, a special briefing should be conducted.
   • Power setting is appropriate for the aircraft configuration and is not below the minimum power for the approach as defined by the operating manual.
   • All briefings and checklists have been conducted.
   • Specific types of approach are stabilized if they also fulfil the following:
     • ILS approaches must be flown within one dot of the glide-slope and localiser.
     • Category II or III approach must be flown within the expanded localiser band.
     • During a circling approach wings should be level on final when the aircraft reaches 300 feet above airport elevation.
     • Unique approach conditions or abnormal conditions requiring a deviation from the above elements of a stabilised approach require a special briefing.

   An approach that becomes unstabilised below 1000 feet above airport elevation in IMC or 500 feet above airport elevation in VMC requires an immediate go-around'.

   Flight Safety Foundation. (2009). Stabilized approach. ALAR briefing note 7.1. Available at https://www.skybrary.aero/bookshelf/books/864.pdf, accessed March 7th, 2020.

   National Transport Safety Board. (2019). Stabilized approaches lead to safe landings. Safety Alert 077, March 2019. Available at https://www.ntsb.gov/safety/safety-alerts/Documents/SA-077.pdf, accessed March 7th, 2020.

   Flight Safety Foundation. (2019). Unstable. Aerosafety World, May 2019. Available at https://flightsafety.org/asw-article/unstable-2/, accessed March 7th, 2020.

   de Boer, Coumou, Hunink and van Bennekom (2014).
7. ... our own processes are intertwined...: Morieux and Tollman (2014) describe that there is a need to *cooperate* to achieve business goals in a complex

environment. They describe the difference of cooperation with *coordination* and *collaboration*. According to them, collaboration is the achievement of 'loose compromises [...] within informal, consensus-seeking groups'. Coordination leads to unnecessary complicatedness by adding dedicated interfaces, structures or procedures. In contrast, cooperation is a difficult social process in which shared goals are realised and overall performance is improved. Each person's success is dependent upon that of others, which is more difficult than either collaboration or co-ordination and will invariably lead to some tension. Too much tension may impede cooperation, and too little may signal that people are able to evade cooperating.

... to establish and uphold descriptions...: According to Hale and Borys (2013a), procedures are a necessary aid in remembering the steps of a complex task (in time), particularly in times of crisis. They form the reference for assessing the way a task is coordinated between actors, executed and documented for future use. Dekker (2014a, 2019, pp. 23–31); Leveson (2011); Hale and Borys (2013b); Hollnagel, Nemeth and Dekker (2009).

8. ... what people have to do to get the job done...: Laurence (2005); Sasangohar, Peres, Williams, Smith and Mannan (2018).

Shorrock (2016) suggests that there are multiple variations of Work-as-Done and Work-as-Imagined. He differentiates between Work-as-Imagined and Work-as-Prescribed, suggesting that the ways that individuals think about work being done varies, and is in incomplete ways captured in regulations, procedures, checklists, standards and job descriptions. Although this is undoubtedly true, for our immediate purposes our previous description of Work-as-Imagined (i.e. Work-as-Prescribed according to Shorrock) suffices. Similarly, Shorrock differentiates between Work-as-Done and Work-as-Disclosed, the latter being the way we write and talk about the work after the fact, tailored to the purpose of the message. It might be closer to either Work-as-Imagined or Work-as-Done depending on the audience. Shorrock is correct in pointing out that to identify Work as (really) Done we need to be close to the front line, observing what is really going on and holding genuine dialogues with front-line workers – rather than 'interviewing' them.

Shorrock. (2016). The varieties of human work. https://humanisticsystems. com/2016/12/05/the-varieties-of-human-work/, accessed August 8th, 2019. See also Berlinger (2016).

9. de Boer, Koncak, Habekotté and Van Hilten (2011).

10. ... it was difficult...: de Boer, Koncak, Habekotté, and Van Hilten (2011).

11. In this study, we compared the actual work practices at four European shunting yards with the procedures. Work-as-Done and Work-as-Imagined were analysed to identify *flaws* (a term to indicate that there are systemic safety shortcomings in an organisation according to Leveson's (2011) STAMP approach). Out of more than 100 data points, four flaws were identified for both Work-as-Done and Work-as-Imagined. Four other flaws were identified for Work-as-Done but not Work-as-Imagined, and three further flaws were found for Work-as-Imagined and not for Work-as-Done. Boelhouwer (2016). Het uitbreiden van STAMP met Work-as-Done, BSc thesis, Aviation Academy, Amsterdam University of Applied Sciences, 2016 [in Dutch].

12. ... a survey amongst aviation companies...: Kaspers, Karanikas, Piric, van Aalst, de Boer and Roelen (2017).

... high-impact low-frequency events...: abbreviated to HILF events, or HILP (high impact low probability). Refers to rare but catastrophic events that are not usually represented in the experience of employees or managers, such as explosions or aircraft crashes.

13. Oil & gas industry, verbal debrief, 2018. The company has since aligned its work procedures better with how work is actually done using micro-experiments. See paragraph 4.3.

14. ... not really visible in superficial observations...: Patterson and Wears (2015)

15. Oil & gas industry, own experience, 2018.

16. ... similar situations...: The focus on routine work rather than on incidents has colloquially been dubbed Safety II:

    Safety management should therefore move from ensuring that 'as few things as possible go wrong' to ensuring that 'as many things as possible go right'. We call this perspective Safety-II; it relates to the system's ability to succeed under varying conditions.

    In: Hollnagel, Wears and Braithwaite (2015, p. 4).

17. Catharina hospital. (2020). Blood test FAQ. V.14, January 2020, https://www.catharinaziekenhuis.nl/paginas/1250-bloedonderzoek-faq.html, accessed January 19th, 2020.

    NOS. (2020). Eindhoven's hospital is recalling 650 patients because of a dirty instrument. https://nos.nl/artikel/2318643-eindhovens-ziekenhuis-roept-650-patienten-terug-vanwege-vies-instrument.html, accessed January 19th, 2020.

18. Air Accident Investigation Unit (2009).

19. Nieuwuur. (2019). Alkmaar never checked AZ stadium's construction, but did sign for it. https://nos.nl/nieuwsuur/artikel/2300887-alkmaar-controleerde-constructie-az-stadion-nooit-maar-tekende-wel-voor-akkoord.html, accessed September 9th, 2019.

20. ... experts come to you...: Guided interactions of managers and leaders with field experts to understand how people create successful outcomes are called 'Appreciative Investigations'.

    ... As external consultants / academics...: Oil industry, own experience, 2018. Aviation maintenance company, own experience, 2019. Hegde, S. (2020). Learning from everyday work. Blog for the Resilience Engineering Association, April 5th, 2020. https://www.resilience-engineering-association.org/blog/2020/04/05/learning-from-everyday-work/, accessed April 17th, 2020.

    ... to ask for narratives...: e.g. Storyconnect, https://storyconnect.nl. The system probes for examples of where work was difficult and through a number of follow-on questions is able to categorize the example. We applied such a strategy successfully at a helicopter maintenance facility, asking for 'a situation you would like to share'. We were able to highlight the large size of the gap between Work-as-Imagined and Work-as-Done. Going undercover might help as well, as the television programme suggests, but we have no first-hand experience in that. Kurtz (2014); de Boer (2016).

    ... balance between efficiency and thoroughness...: Hollnagel (2009); Amalberti and Vincent (2020).

21. The list of questions is from: Hummerdale. (2015). If it ain't broke, fix it anyway. Blog on http://www.safetydifferently.com/if-it-aint-broke-fix-it-anyway/, dated February 25th, 2015, accessed August 18th, 2019.

22. ... expect your direct reports...: Morieux and Tollman (2014, p. 171).

23. ... lowest level (self-)audits...: One company was able to increase what is learned from audits by focusing on curiosity, not compliance, and quality, not quantity. Discussed in paragraph 4.3 as an example of a micro-experiment.

24. ... to involve the people...: The need to involve users in the design of rules and procedures is akin to the move towards participatory design, in which end-users are involved in the design process of artefacts that influence their work and life.
    Tromp and Hekkert (2018, p. 12).

25. Oil industry, own experience, 2018.

26. ... Our approach assumes...: Much of our approach regarding trust, participation and decentralisation reflects Alex Pentland's latest thinking as is documented in his new book titled *Building the New Economy* (available from Fall 2020 at MIT Press), at least as reflected in a recent interview with a Dutch national paper.
    Van Noort, W. De Wereld heeft een nieuw besturingssysteem nodig, June 20th, 2020, available at https://www.nrc.nl/nieuws/2020/06/19/de-wereld-heeft-een-nieuw-besturingssysteem-nodig-a4003444, accessed June 21st, 2020.

27. ... We want the gap...: Hale and Borys (2013b).
    ... On the one hand...: Hale and Borys (2013a).

28. Compared to the original (Hale & Borys 2013b), I have deleted superfluous steps and normative language that focuses too much attention on individual performance rather than systemic factors. It is now possible to discern a 'use loop' (consisting of steps 1, 2, 7 and 8) and a 'design loop' (steps 3–6) in the process. The proposed rule management process will reconsider rules when triggered by snags in practice, rather than a reappraisal after a predetermined period such as is sometimes advocated but usurps scare resources. It seems that most organisations have better processes for creating rules than binning obsolete ones. We sometimes see 'rules' are made locally with the best intentions but with little concurrence within the wider organisation about their status. Needless to say, the 'read and sign' procedure for communicating new or modified rules is absolutely contrary to the idea of an 'appropriate implementation'.

29. ... What we would like to achieve...: That operators feel trusted to make the right choices when they are executing their tasks, and the need to deviate from the procedures is signalled to peers or supervisors is sometimes called 'freedom-in-a-frame'.
    See Shorrock, Wennerberg and Licu (2018).

30. ... the adherence to rules is...: Ruitenburg (2017).
    ... measures to counter speeding violations...: Automatic detection of traffic violations can be very effective. One study reports a reduction from 24% to 2.3% of the drivers that were speeding or passing through a red light. Sagberg (2000).
    ... garbage...: See Ruitenburg (2017).

31. ... formalized processes such as audits...: We generally differentiate between internal audits and external audits by regulators, certification bodies and supply chain partners. There is limited freedom in accepting external audits from certification bodies and supply chain because of the commercial consequences. Internal audits have an objective to learn and assure our processes and should be treated as such. Initially, these will focus on understanding reality (see Section 3.4), and later, these will help maintain alignment between *paper* and *practice* (3.6). External audits are aimed at assuring that we have achieved alignment and can be used to identify gaps for which the organisation is blind. I have found that once alignment has largely been achieved, the burden of audits diminishes and their value increases.

... the adherence to rules is...: This type of working environment exceeds that of a just or reporting culture and extends into all facets of Reason's (1998) safety culture: informing, learning and flexible. Although it is not possible to directly improve a culture, we have found that certain actionable prerequisites can trigger a movement in the right direction. These include a confidential, non-punitive reporting system, freely accessible safety information, sharing safety information across departments, etc. Piric, Roelen, Karanikas, Kaspers, van Aalst and de Boer (2018); Karanikas, Soltani, de Boer and Roelen (2016); Reason (1998).

... the organisation (not the auditors!) decide...: Achieving parity with auditors requires the organisation to be fully cognisant of the gaps between Work-as-Imagined and Work-as-Done, so that there are no surprising findings. The exceptions that will be found can then be justified. I have found that it is permissible to agree to disagree with auditors on a limited number of minor findings, providing alignment has largely been achieved.

32. Internal email, confidential, January 2020.
33. ... a significant shift...: Dare I use the word: 'cultural'?

    ... a generic guide...: Service, Hallsworth, Halpern, Algate, Gallagher, Nguyen, Ruda and Sanders (2014); Shorrock (2020).
34. ... a physical infrastructure...: one major bottleneck in involving employees as much as recommended in our approach is workload. Some improvements have been delayed for several months because the field had 'other priorities'. Oil & gas industry, 2018, own experiences.

4

Effective Workplace Innovations

4.1 Introduction

By executing the steps explained in the previous chapter, we will have generated procedures that are as close to how work is actually done as reasonably possible, and where exceptions are identified and acted upon. This leads to a reasonably stable set of rules that will only slowly adapt to changes in the environment. Whenever we need to amend rules and procedures more rapidly, want to innovate or need to close a large gap between *paper* and *practice*, a different approach is required. In a complex environment, where outcomes are difficult to predict, we cannot be sure that the interventions that we devise actually bring the results that we envisage, or that unintended side effects are avoided. We need an approach that allows us to try different things, does not cause peril and can be stopped if we don't get the results that we want. This is where the micro-experiments come in that I will introduce in this chapter. But firstly, I will characterise the kind of place that we inhabit. My suggestion, based on years of scientific inquiry and practical experience, is that this planet is actually quite a complex place. The world around us doesn't always behave as we presume.

4.2 Complexity

We have come to realise that the organisations that we are part of are 'complex' – a scientific term meaning that the interactions within an organisation give rise to consequences that are not visible when we consider only isolated parts of the organisation or individual actors. Small causes can lead to exorbitant effects (called the 'cause–effect asymmetry' or the 'butterfly effect'). These characteristics mean that it is impossible to predict exactly what the effect is of any interventions, policy changes or new tools that we implement. We might introduce consequences we hadn't considered beforehand or overshoot our original aim.[1]

Parents used to arrive late to collect their children in a group of day-care centres in Israel, forcing a teacher to stay after closing. This was frowned upon because the teacher was not paid for this extra time. A monetary fine was introduced for late-coming parents to deter this behaviour, as it was felt that social condemnation alone was too ineffective. However, as a result, the number of late-coming parents did not decrease but instead increased significantly. Even worse, after the fine was removed, this behaviour was maintained and there was no decrease back to the original levels in the number of late collections. Apparently, and in hindsight, complex social and economic interactions occur that run counter to the original intent of the intervention by management. The fine was interpreted as a price to pay for the "added service" and the social condemnation by teachers whenever a parent was late was re-evaluated to be less of an impediment than before.[2]

It was impossible to predict the exact effects of any intervention in the complexity of the day-care centre, but we can explain it in retrospect. It is quite probable that a different combination of cost, communication and condemnation would be more effective, and so experimentation with these factors is warranted. Similarly, we have seen examples of unintended effects of incentive schemes (e.g. for lost-time injuries or zero-harm) which prejudiced the short-term results over longer-term improvements. In the complex domain – full of uncertainty and flux – experimentation is necessary to identify the patterns that emerge and see which interventions are effective, and which aren't. We should carefully monitor the effect of any intervention that we contemplate. An approach is required in which trying things out is permissible and the results are tracked.[3]

4.3 Micro-experiments

A proven way to try things out while avoiding the risk of considerable problems is to use micro-experiments. Micro-experiments are safe-to-fail interventions that are monitored for success or failure.

The Woolworth retail chain in Australia together with safety scientists executed micro-experiments in an effort to create more bottom-up safety. The shops were divided into three groups, two of which were allowed to define their own safety rules. Shops in one of these groups were facilitated and coached, the other group was left to their own devises. The third and final group was left unchanged to compare against as a control group. It turned out that the first group (free but facilitated) was more effective in creating new, effective safety rules and avoiding adverse events than the other two groups. They did this by trying out different things, some of which were modified or discarded. Other interventions turned out to be successful and were retained. The second group (free, not facilitated) also tried things out but were less effective in choosing what to continue. Both groups were better than the third (control) group that represented the original context.[4]

Micro-experiments are different to ad hoc trial-and-error interventions in that they are planned before they are executed, and carefully monitored to see whether the intended results are realised. They are different to large-scale projects in that those are not supposed to fail, whereas micro-experiments are temporary and we allow for unexpected, even disappointing results.[5]

> In a control room we once found that operators coped with over 100 open alarms – filtering the important ones from the trivial ones by common sense rather than any additional support. The Safety Officer of the company that accompanied us was taken aback: hadn't they spent lots of effort in making an advisory database for these alarms? The control room operator reluctantly conceded that they didn't use it because it was so cumbersome. "Well, show me", said the Safety Officer. It took quite a while to load the database, but eventually it was up and running. "Ok, let's use it on this first alarm", suggested the manager. So, the operator looked at it, and said: "Well no, this suggestion is outdated – we have invented a better solution recently that is not yet included". At that point I suggested that this database needed to be revisited, possibly from scratch. But the Safety Officer was adamant that he could not do that: "Do you know how much effort has gone into this project already?!"[6]

In this example, the project had grown too large to be allowed to fail. But in a complex environment, we cannot have the (misplaced) confidence that any intervention will surely deliver the predefined results as expected. We need to allow for the possibility for the intervention not to work. In fact, the experiment can be considered a success if we find that an intervention 'fails' and that some other solution is required. We therefore need to be able to stop or slow the experiments. On the other hand, if all goes well, then the intervention is ready to be formally implemented and possibly enlarged on a wider (geographic or organisational) scale. Naturally, we will create the ideas for the micro-experiments in close collaboration with the task executors; many ideas will derive from them. Micro-experiments generally require temporary safeguards because we will be deviating from the existing processes, perhaps violating the constraints that are usually in place. Care needs to be taken to identify and only maintain those constraints that are absolutely vital to safeguard against unacceptable risks – we value the input of safety experts here. The criteria by which we determine whether an intervention is successful (or not) need to be determined ahead of time, although additional insights will occur as experience is gathered. Interventions that are deemed successful need to be incorporated in the next version of Work-as-Imagined. Even after formalisation, we remain vigilant about unintended effects, expecting the front line to report where exceptions are necessary. Multiple micro-experiments can run in parallel, preferably somewhat isolated from each other.

4.4 Designing and Executing Micro-experiments

Good targets for micro-experiments are procedures that are not delivering results as expected. Many organisations that we first engage with are susceptible to large and potentially embarrassing gaps between Work-as-Done and Work-as-Imagined, which are not straightforward to close. This is often a good starting point to try out micro-experiments. Other reasons to execute micro-experiments include adapting rules to changes in processes or equipment, or for innovation.

To further define the problem, we first need to identify the 'job to be done': what operators really seek to accomplish in the given circumstances. Successful micro-experiments help make the job safer and easier. Your first initiatives might be somewhat top-down, but we would encourage you to acknowledge bottom-up initiatives as experience grows. This may take some getting used to by management and front line.[7]

Some examples of where micro-experiments have been developed are as follows:[8]

- Improvement of procedures on a gas rig. The existing procedures were too cumbersome to be useful. New procedures based on the way that work was actually done were written by the front line, and input was given by technical process experts. The new procedures were first tested in practice and optimised before they were formalised.

- Increasing what is learned from audits by focusing on curiosity, not compliance, and quality, not quantity. The number of audits was halved to make room for a more in-depth inquiry, and the name was changed to 'learning review'. Questions that reflected curiosity were added, such as 'What is your biggest frustration about this process?', 'What is the stupidest thing we make you do?' and 'When was work successful yet difficult?'

- A ship's captain is expected to stay on the bridge during the passing of the Suez Canal from the Bay of Aden right up to the Mediterranean Sea – a total of more than 50 hours. How can we ensure that the master is not too fatigued during and after this passage? Interventions considered include resting a few minutes at a time, placing a second master on board for the passage and delegating of tasks to the first officer or other crew members. Because of the number of ships owned by the company, different ideas could be tried within a few months.

- Several micro-experiments were developed for an aviation maintenance facility, including the introduction of seat covers to protect business class seats after they were installed, removal of fuel fumes under the wing of aircraft and the use of new suction cups as a restraint when working at heights on the fuselage or wings of the aircraft. Several options were tested and then the best were formally implemented.

- Following a fatal accident, yellow and black road markings were added to a bridge on the moving parts of the deck. These were tested to see whether they were effective in nudging cyclists and pedestrians to avoid stopping on the incorrect side of the barriers, where they risk being lifted as the bridge opens for boats.

It seems to be beneficial to start micro-experiments as small as possible, within a small department or sub-organisation and initially with interventions that target recognisable differences between Work-as-Done and Work-as-Imagined – gaps that we all know are there and are hard to tackle. At times, we have found quite a bit of resistance with lower managers to these structured interventions, as they prefer a more ad hoc approach. Quality and other support staff, on the other hand, prefer to add so many constraints that the experiment becomes unworkable. The solution is to keep things small and time-constrained, and create a sense of urgency that pushes the experiment ahead.[9]

Key to a good micro-experiment is to devise a great number of far-reaching interventions before settling on a few that seem the most promising to actually try. This brainstorm allows you to widen the scope of possible solutions and therefore get a better view of what might or might not work. It is more collaborative and fun too. After we have a good long list, it usually starts to emerge which ones we would like to try first, based on potential and effort. In case there are too many suggestions, we suggest that a cost–benefit trade-off is used to set priorities. Some experiments will not be possible because of a lack of resources or because too many risks are introduced. Depending on the situation, we might be able to try different things in parallel. Initially, it is probably best to limit the number of micro-experiments to be initiated and running in parallel to about three to five. As the organisation becomes more familiar with micro-experiments, this constraint can be relaxed.[10]

To demonstrate the effectiveness of the intervention, it is a good idea to compare it with a no-change situation. The attention and training associated with the micro-experiment may be part of the reason that the intervention seems successful, rather than the intervention itself. In some cases, it may only be possible to compare the result with some historical baseline (e.g. last month's or last year's data), while in other cases, several interventions will run in parallel with the original situation. Results to compare across conditions include the intended ones, but also unexpected side effects and the subjective judgement of those involved.[11]

An important consideration is how to make sure that the micro-experiment doesn't introduce intolerable risks and is acceptable for management and regulatory authorities. In some cases, it took 18 months to get sufficient buy-in to embark on micro-experiments on a large scale, but in other cases, the buy-in was immediate because safety professionals were on board with the idea and the initial experiments were low risk. Buy-in is easier if you target a recognisable gap between Work-as-Done and Work-as-Imagined, start as small as possible, ensure sufficient additional safeguards are in place and involve those with responsibility for oversight from the start. [12]

I recommend determining beforehand on what basis the intervention is deemed effective. What effect are we trying to achieve and how do we make these results tangible? Which degree of improvement signifies success relative to the baseline case? What side effects should we be wary of (accepting that we cannot predict all – emergent – behaviours)? Many micro-experiments include qualitative or subjective measures to determine success, such as 'How easy was it to operate?' and 'What did we like about it?' It is important to have regular reviews about the progress of each micro-experiment and the achievement (or not) of the criteria. This allows us to actively and collectively maintain, amplify or dampen the micro-experiment, or even end it prematurely. The reviews should be scheduled ahead of time (for instance, as part of a regular cycle of operational meetings). A monthly or quarterly or so report to upper management seems advisable to keep them in the loop.

An essential part of the design of micro-experiments is deciding up-front how to stop the experiment when the intervention does not generate sufficient positive results, or the side effects are too detrimental. In most cases, this will mean discontinuing the intervention and returning to the original situation but watch out: not everything within a complex system is reversible – you may have unleashed powers that are hard to contain. Remember the day-care centre in Israel that tried to penalise parents for coming late to collect their child? When that didn't work and the penalties were withdrawn – supposedly reverting to the original situation – parents continued to come late. We also need to consider beforehand how we will increase the scope of the intervention if it turns out to be successful. Generally, the intervention can be expanded across multiple installations, production lines or entities. But if our intervention requires many resources that were only available for an experiment but not on a regular basis, it will be difficult to scale up. In that case, additional experiments may be required to achieve similar effects with less resources.[13]

The micro-experiment should be ended at a predetermined time, even if it is running well. This is the right moment to reflect on the cost and usefulness of the intervention and consider whether it will form part of the standard repertoire moving forward. Once we have concluded that we want to maintain an intervention, we need to end the experimentation stage and formalise the procedure to make it routine. How to go about this has been described in Section 3.6.

4.5 Conclusion

A micro-experiment is an effective method to test workplace innovations that are better suited to complex environments than ad hoc interventions or large-scale projects. They allow us to generate opportunities to innovate our processes and maintain alignment with the needs of the business in a controlled manner. Micro-experiments are safe to fail, and we are able to dampen and amplify them contingent to meeting criteria which we have determined ahead of time. Interventions that are deemed successful will be formalised. To successfully execute micro-experiments, a leadership style is required that acknowledges complexity and that allows for controlled experimentation. It differs from the traditional command-and-control management styles: these fail as managers become impatient and interventions fail to achieve the predetermined results.

Key actions:
- Choose micro-experiments over ad hoc interventions in complex environments to track benefits, and ensure that these are fully realised.
- Choose micro-experiments over large-scale projects in complex environments to trial multiple solutions and accommodate unexpected results.
- Define the problem to be tackled, and identify multiple solutions. Choose interventions with the most promising cost–benefit ratio.
- Design the micro-experiment:
  - Keep small
  - Put constraints in place to limit new risks
  - Decide on success criteria
  - Identify how to stop or enlarge the intervention
  - Set a fixed duration
  - Identify a baseline to compare results with.
- Execute, monitor and terminate as per the plan.
- Decide on subsequent steps: formalise, modify or abandon.

It would seem that having tackled the – at times challenging – steps that have been explained in the last two chapters that we can now rest on our laurels and assume we have achieved safety leadership. Alas, as we will see in the next chapter, any organisation is subject to the ubiquity of eroding safety margins, and so there is more to do.

Notes

1. ... complexity...: Traditional western thinking assumes a Newtonian–Cartesian reality – that is what we have been taught in school. This world view is based on an unbridgeable gap between the human mind and physical world (Descartes), the latter defined by unequivocal space–time coordinates and governed by universal physical laws (Newton). The Newtonian–Cartesian paradigm permits the decomposition of any system into smaller parts without the

loss of predictability. Components are isolated, separate, interchangeable and unchanging. Descartes and Newton, as representatives of the scientific revolution, strived for judgements that are unconditionally correct. An organisation founded on Newtonian–Cartesian (or mechanistic) principles aims for hierarchy and control through a central bureaucracy, fixed roles and unambiguity. In a Newtonian–Cartesian world, complete knowledge is possible, and harm is (therefore) foreseeable. Effects have proportional causes (i.e. big mistakes lead to large accidents, and small mistakes to incidents). Actors are judged absolutely and either perform appropriately (i.e. according to the mandated rules) or fail. Time is reversible: a sequence of events leading up to an accident can be traced back to identify the 'failures'. As we now know, the Newtonian–Cartesian paradigm is a useful model but only partly correct. System behaviour cannot always be reduced to component behaviour: the interrelationship between components and actors in a socio-technical system gives rise to 'emergent' behaviour that is not visible when the components or actors are isolated from each other (see glossary for an exacter description). Small causes can lead to exorbitant effects (called the 'cause–effect asymmetry' or 'the butterfly effect'). Complete knowledge of a complex socio-technical system is impossible because of the open nature of the system and its resultant sensitivity to triggers from the environment. Complex systems demonstrate history or path dependence: it matters how a certain situation was arrived at.

   Zohar and Marshall (1994); Dekker, Cilliers and Hofmeyr (2011); Morieux and Tollman (2014).
2. Gneezy and Rustichini (2000).
3. … unintended effects of incentive schemes…: For the discussion on incentive schemes for lost-time injuries and zero-harm targets, see Chapter 2.
   … In the complex domain…: Complexity and the idea behind micro-experiments are described well in Snowden and Boone (2007). An interesting application in the South African mining industry is described in Deloitte (2009).
4. Dekker, S. (2018). The Woolworth experiment. https://safetydifferently.com/the-woolworths-experiment/, accessed January 22nd, 2020.
5. Micro-experiments mirror Design Thinking: they are also a way of approaching organisational challenges with a user-first mindset before designing and testing solutions quickly in order to see effective results. The [...] process is flexible and non-linear in nature, allowing teams to go back and forth between ideation, testing, and user definition as best suits the project. You might learn things from testing that allow you to better ideate and understand your users.
   https://www.sessionlab.com/blog/design-thinking-online-tools/#design-thinking-faq, accessed June 16th, 2020.
6. Oil industry, own experience, 2018.
7. … 'job to be done'…: This thinking is an adaption of Christensen's theory of Jobs to Be Done for consumer marketing. He suggests that people don't simply buy products or services; they pull them into their lives to advance in one way or the other. He calls this progress the 'job' they are trying to get done: what an individual really seeks to accomplish in a given context. These circumstances (in consumer marketing) are more important than customer characteristics, product attributes, new technologies or trends. Good innovations solve problems that formerly had only inadequate solutions – or no solution at all. Jobs are

never simply about function – they also have powerful social and emotional dimensions. This tenet seems to be equally valid in the safety domain with a few adaptations that are of my own making. In safety, whether an operator buys into safety rules is determined by the headway that he or she is trying to make in specific circumstances, not their characteristics or (just) the rule itself. Christensen, Hall, Dillon and Duncan (2016).

8. https://www.youtube.com/watch?v=8WymIJ84ISM&feature=youtu.be, accessed June 26th, 2020.

   … procedures on a gas rig…: The procedure for changing the filter was improved in a workshop and then implemented as a micro-experiment to test its suitability. The intervention is deemed successful and has been formalised.

   … what is learned from audits…: Changes to the self-audits in the oil & gas industry were implemented as micro-experiments. The intervention is deemed successful and has been formalised.

   … Aship's captain…: The ideas for micro-experiments on the ships' manning levels were developed in a master class on Human Factors and Safety, Amsterdam, January 21st–25th, 2019. It is unclear whether the micro-experiments have been executed.

   … Several micro-experiments…: Several of the micro-experiment in the maintenance environment have been implemented. Plioutsias, Stamoulis, Papanikou and de Boer (2020).

   … Improving the design of a road bridge…: Caprari and Van Wincoop (2020)

   Dutch Safety Board. (2016). Accident at Den Uyl Bridge, Zaandam. https://www.onderzoeksraad.nl/en/page/3748/accident-at-den-uyl-bridge-zaandam, accessed June 26th, 2020.

   Although these examples stem from commercial organisations, micro-experiments are equally justified in the public sector. See, for instance, Taylor, B. (2017). Benjamin Taylor – The Future of Commissioning and Transformation – Public Sector Show from 17:40 minutes.

9. … add so many constraints…: The constraints that were initially imposed at one company and later turned out to be unworkable included:

- Any actions must be according to the regulations.
- Any intervention must be according to the procedures.
- Any intervention cannot create any new risks.
- Any intervention is according to the company's risk assessment.
- Any intervention is not a permanent solution.
- Any intervention must comply with requirements set by [company's] clients.
- Any intervention is only applied to the predetermined bay with a predetermined group.
- Any intervention will run for a fixed period of time.
- Any intervention can only be executed with the presence of at least one of the [researchers] present, on site.
- Any intervention can only be executed in agreement with QM, BP, the bay manager involved and the project team.

See Beckers (2019); Plioutsias, Stamoulis, Papanikou and de Boer (2020).

10. The design-thinking process can be used to support the success of micro-experiments. 'Design thinking is a human-centred process of finding creative and innovative solutions to problems' that has been tried and tested by designers and is now being rolled out in organisations much more broadly.

    By approaching the process using design methods, organisations and teams in any field can better understand their users, redefine challenges, and quickly test and iterate on possible solutions. [...] The design thinking process has five stages that can be approached in both a linear and non-linear fashion. The five stages of Design Thinking are: Empathize, Define Ideate, Prototype, and Test. https://www.sessionlab.com/blog/design-thinking-online-tools/#design-thinking-faq, accessed June 16th, 2020.

11. ... to compare it with a no-change situation...: Often called the 'control' group, for instance, as applied in the Woolworth case. Dekker (2018).

12. ... it took 18 months...: Dekker (2018).

13. ... Remember the day-care centre...: Gneezy and Rustichini (2000).

5

Staying Safe

5.1 Introduction

It stands to reason that where we have multiple objectives to fulfil, we continuously need to balance efficiency and thoroughness. As your people successfully navigate the precipitous cliffs of daily life to achieve their goals, they discover what works and out of habit or design this becomes the new standard. The success of these adaptations masks what is really going on, resulting in a system that is operating in a riskier state than you might consciously choose. Adverse events are relatively rare, and therefore there is a natural bias to underestimate the effect of these small modifications on safety. The gradual tendency to erode safety margins and accept (knowingly or unknowingly) more risky behaviour is an organisational phenomenon sometimes called 'drift' – as in 'drifting into failure'. Drift can be countered by promoting discussions about risk and building expertise. But like the lunch, both of these don't come free. So it is all about finding the right balance between safety and other goals. In this chapter, we first describe the factors of drift and how to identify them, and we then give suggestions how to counter the subsequent erosion of safety margins.[1]

5.2 Drifting into Failure

The erosion of safety margins is a naturally occurring organisational phenomenon that is difficult to avert and cannot be attributed to individuals. Drift is fuelled – in the absence of significant incidents – by five factors: scarcity and competition, decrementalism, sensitivity to initial conditions, unruly technology and failing protective structures. We discuss each of these to show how to identify drift and emphasise the inevitability of the erosion of safety margins within organisations.[2]

5.2.1 Competition and Scarcity

All organisations operate in an environment of competition and scarcity of resources. As a consequence, safety cannot be the only objective of the organisation; other goals like being profitable enough for continuity, attracting and retaining employees and creating value for customers also need to be satisfied. Even monopolists have to take regulators and possible substitutes into account, and governmental agencies are subject to austerity. In businesses, competition and scarcity of resources as a factor in drift are often visible through a predominance of financial and productivity goals which overshadow all other objectives, including safety.[3]

> A vivid example of how the balance can be lost between economic goals and safety goals is shown by speeches given by the CEO's of British Petroleum in the period just before the Deepwater Horizon disaster in 2010. For instance, in the Annual General Meeting of Shareholders five days before the accident the then CEO Tony Hayward declares that "Our priorities which lie at the heart of all our operations remain safety, people and performance". But in fact, in much of this speech he embraces economic efficiency and financial performance rather than safety. The text is dominated by statements concerning financial matters and organisational efficiency whereas the word 'safety' is used only once more in the entire speech.[4]

The scarcity of resources as a factor in drift are caused by economic pressures, and may take a while to surface:

> On 21 August 2017, a fire started around 9:30 pm in the ExxonMobil refinery in Rotterdam, the Netherlands. This fire took place in a furnace of the Powerformer factory. Before the fire started a large disruption had occurred, which led to emergency shutdowns of all six stoves in this plant. When all six furnaces were restarted simultaneously, there was no fluid flow leading to overheating of the heating spiral in one of them. It ruptured and the flammable process fluid came into contact with the heating elements and ignited. The no-flow alarm was overlooked by the operator in the frenzy of alarms that occurred. This is not surprising as the hot start of the six furnaces relied on 23 (!) sensor bypasses and generated 250 alarms in ten minutes, a quarter of which were of the highest category. These overrides had become normalised over the course of several preceding years. The Dutch Safety Board concludes that it was impossible for just one operator (as was the case) to monitor all these alarms and act on them.[5]

Both the bypassing of the sensors and the unacceptable high number of alarms during certain phases of operation were known within the company and were not acted upon – a sure symptom of drift. Scarcity in resources might not just be in number, but also in expertise. This seems to have played a role in the running aground of the Costa Concordia in January 2012. Due to the growth in demand for vacation cruises, ships' officers were promoted at an unhealthy rate with limited experience under their belts. In the case of the Costa Concordia, the ship's captain was – amidst other issues – not able to rely on his deck officers to identify the lag in the turn away from the island of Giglio to avoid the impending disaster.[6]

In the public sector, scarcity is often encountered when the management agenda is dominated by mission criticality or compliance with bureaucracy, particularly in times of austerity:

> At a university in England the quality of any degree programme is assured through a number of measures, such as: peer review of the assessments prior to these being presented to the students, evaluations of each module, discussions with students about the entire programme each semester, discussions with industry representatives, and consultation with external examiners from outside the university. Each of these need to be written up and then collected in an annual programme review, including identifying and monitoring improvement points. The burden of the large number of actions, impeded by cumbersome IT systems, is so high, that the responsible academic was heard to sigh: "we understand the importance of each action and so we'll see how far down the list we get in actioning each of these". In other words, it was impossible to get everything done and so available time and personal priorities would determine which actions were undertaken and which were ignored.[7]

When resources are stretched, people make their own choices in setting priorities. It seems to make much more sense to lower the administrative burden and/or increase resources so that the things that the organisation considers to be important get done, rather than leaving it to personal choice. Cases like this occur where additional well-intended measures clash with pressures to reduce staffing levels and maintain output. In many ways, it is completely logical that there is downward pressure on staffing levels, and it is difficult to judge to what extent this is acceptable – except in hindsight after an adverse event...[8]

5.2.2 Decrementalism

A second symptom of drift is decrementalism: the tendency to judge risks as more acceptable if the hazard is introduced in small steps rather than one big change.

> On a gas rig in the North Sea we were talking with the control room operators. Over time, the staffing of the control room was reduced from four operators, to three operators, to two operators and a trainee – without substantial reductions in workload, in fact to the contrary. As a consequence, it was no longer possible to execute tasks out on the rig with two experienced operators as was sometimes felt necessary while leaving the control room properly attended.[9]

Elements of decrementalism are visible when people within an organisation become so accustomed to deviant behaviour that they don't consider it as illicit anymore.[10]

5.2.3 Sensitivity to Initial Conditions

The third factor of drift is the sensitivity of safety to initial conditions: whether the initial design and implementation of the system is still adequate, as conditions and requirements might have been changed over time and no longer meet our current needs for safety.

> In June 2014, two large explosions and a fire occurred at a chemical process plant in Moerdijk, the Netherlands during the start-up of the process. Luckily, there were only limited injuries to two operators working nearby. The plant itself was destroyed and debris was scattered up to 800 meters away. Nearly forty years earlier the company had determined that the priming substances in the reactor (a liquid and a catalyst) did not chemically react. However, over time a more potent variant of the catalyst was chosen, and therefore the assumptions about the process that had been made during the design of the plant were no longer valid. Despite these and other modifications to the manufacturing plant and the processes, no new risk analysis was performed.[11]

This example shows the sensitivity of safety to initial conditions, and the need to revisit these regularly during the lifetime of the asset to determine whether the assumptions that were made in the past are still valid.[12]

5.2.4 Unruly Technology

The fourth factor in drift is unruly technology. This is technology (often of the information-processing kind) that conforms to specifications yet in practice displays unexpected behaviour. The automation is not malfunctioning, but the automation logic does not match the operator's sense of what should be happening. This makes the automation untrustworthy from the user's perspective because its inner workings are not understood.[13]

> The Boeing 737 that crashed just before landing at Amsterdam Airport in February 2009 had a faulty radio-altimeter on the left-hand (captain's) side. It was .stuck at an elevation of 8 meters (24 feet), suggesting that the aircraft was close to the ground. This was known to the pilots, but what they did not know was that it was connected to the right-hand auto throttle, not the left-hand auto throttle as one might expect. There are multiple types of altimeters in an aircraft for different flight phases. The right-hand auto throttle was engaged in the approach, and as the landing configuration was established, the auto throttle began to take its input from the faulty left-hand radio altimeter. The crew did not realise that the auto-throttle was spinning the engines down to idle as would be the case on landing while they were still several hundred meters up in the air. Before they understood what was happening and could alleviate the situation, the aircraft stalled and fell to the ground.[14]

'Automation Surprises' occur when operators are surprised by actions taken (or not taken) by the automated system. They are a relatively frequent phenomenon and occur once every 3 months for an average airline pilot, and similar high occurrence rates are typical in other industries. In a control room, we found that operators coped with over hundred open alarms – filtering the important ones from the trivial ones by common sense rather than any additional support. Often these symptoms of unruly technology are without significant consequences – but they generate a wealth of opportunities to identify areas for improvement. Hopefully, it will be possible to change the system logic to better align with operators' expectations, but failing that, it seems prudent to at least share experiences and solutions among operators.[15]

5.2.5 Contribution of Protective Structures

The fifth and final factor in drift is the contribution of protective structures like auditors, government authorities and safety professionals. These protective structures are intended to independently evaluate and challenge the safety of the system, but in the absence of adverse events, there is a natural tendency to subscribe to the status quo and not be seen as a nuisance. The very entities that are supposed to oppose drift might make the (mal)functioning of the system opaque. Social relations, economic bonds and career opportunities of an in-house department quickly defy the rational need for independent wariness.[16]

> In October 2018 and March 2019, the relatively new Boeing 787 Max was involved in two separate crashes, killing 346 people. After the second crash, the head of the American aviation regulator (the Federal Aviation Administration, FAA) conceded that "his agency had made mistakes in its handling of the crashes". An FAA analysis after the first crash off the coast of Indonesia determined that the Max was likely to crash again if regulators did not act. The FAA did not ground the aircraft pending modifications by Boeing, relying instead on enhanced instructions to pilots. A second plane then crashed under similar circumstances in Ethiopia five months later. A government committee after the second crash suggested that "FAA managers sided with Boeing instead of their own safety experts." The FAA kept the Max flying for several days after the second crash, despite grounding by European and numerous national aviation regulators. In internal communications, Boeing employees dismiss the Federal Aviation Administration and describe regulators as "dogs watching TV." Another time, a Boeing employee wrote: "There is no confidence that the FAA is understanding what they are accepting (or rejecting)." [17]

Examples of the contribution of internal protective structures to drift include situations where the safety department was working with the operational departments to achieve the planned number of audits; to monitor the execution of action items independent of their urgency or importance; where there was no opportunity to have a discussion about which mitigations are trivial or significant; and where internal audits were used to demonstrate compliance instead of being used as an opportunity for learning.

An oil and gas plant is required to execute audits locally that are aimed at self-assurance and learning – by the business for the business. They are complemented by audits that are executed by corporate staff and external audits by or on behalf of the regulator. However, in practice, the local audits were not directed at learning at all. The introductory text of the audit document states: "To ensure [company] is complying with the legislation, and in accordance with the [regulator's] guide, you are required to complete the following question set." The questions ask about compliance with regulations, not about how the regulations do or do not support the task at hand. In one specific case the audit did not signal deviations between the prescribed isolation of process equipment and current practice, which was less rigorous than engineering had recommended. The (permanent) exemption for this was granted locally, without involving engineering even for a retrospective judgment, thus circumventing their expertise. This was never picked up by any of the internal auditors who considered this practice normal.[18]

Remember also the earlier example (paragraph 3.7) of the inspection by the fire department, and the request to remove obstructions exclusively for the purpose of the audit. This is another example of a protective structure working with the organisation to achieve the short-term goal of fewer findings and no embarrassment, rather than using an external audit to challenge the status quo and improve safety.[19]

5.3 Countering Eroding Safety Margins

Safety margins need to be defended against erosion. In this section, I will discuss how to counter the effects of drift into failure. All organisations are susceptible to the erosion of safety margins, but some organisations are better than others in sustaining their operations in the face of adverse or unexpected conditions. The absence of a major incident should never be taken as a sign that you are on the right track. A complex system is subject to delays in responses and emergent behaviour, and its conduct is sensitive to specific configurations that appear sporadically or may develop over time. Central to safeguarding against drift are two measures: keeping the discussion on risk alive and thereby striking a balance between safety concerns and other organisational goals, and allowing operators to build their expertise in a safe-to-fail manner. I will discuss each in turn.[20]

5.3.1 Keeping the Discussion on Risk Alive

We need to ensure that discussions about staffing levels, maintenance backlogs, technology and innovations are not limited to returns on investment and time to market but also include safety, risks and emergent behaviour.

This is easier read than done, and from the outside, it is difficult to discern whether a balance has been struck until the metaphorical dreck hits the fan.

Internally however, several characteristics of a lively discussion on risks should be visible. These include dissent, delays, diversity and deference to expertise. They all need psychological safety (discussed earlier) to succeed. Diversity of opinion and the possibility to voice dissent is important to allow real discussions about safety to take place. The discussions on risk need to be real and animated, at all levels, from board room to toolbox, with conclusions that can vary depending on the situation. The outcome of these discussions can include stopping operations, or adding additional assurances or resources, or delaying actions until the circumstances are more favourable – options that are not always appreciated by 'successful' business school managers. There are no right answers, and an equilibrium between safety and other goals needs to be found anew in every instance.[21]

Differences in professional backgrounds and gender or ethnic diversity are ways of creating a variety of perspectives that are conducive to more lively discussions. A greater diversity in many organisations can also be achieved by making use of the (technical) expertise that is already available but sometimes not heard. To protect against drift, those familiar with the messy details of work need to be encouraged to speak up, particularly if their words go against the emerging consensus. Allow those with engineering, operational and organisational expertise to make decisions (rather than deciding for them), and break down barriers between hierarchies and departments, to enable peers across the organisation (and even beyond) to discuss, argue and agree. Initially, these confrontations might not be welcomed by the recipients, and so some drive by you might be required. Facilitate the discussion by ensuring that common goals are maintained on the organisation's mission and values. It is worthwhile to revisit the outcomes of earlier discussions after the event has passed: do we believe we struck the right balance, were we overcautious or were we lucky to get by unscathed? Whenever our operations are going to be impacted by significant changes, it is prudent to require a risk analysis by safety professionals. Through job rotation or comparable means, we can recalibrate our sense of what is 'normal'.[22]

You know you are on the wrong side of the curve if you are window dressing for audits and inspections, if your experts know things they are not sharing with you and if only after adverse events, you are finding out about how work really gets done. Consider the following example of a minister being caught out:

> The Dutch minister for economic affairs was portrayed on television in his limousine without a seat belt. He was being interviewed in the back seat of his car for a current affairs programme and the journalist diligently strapped himself in. The minister indicated with a boyish grin that he often neglected to wear his seat belt when being driven because it was uncomfortable. After broad exposure in the press, several of his fellow ministers indicated that this was irresponsible behaviour. "I am counting

on him never to do this again," muttered the minister for traffic and infrastructure Cora van Nieuwenhuizen. "There are still too many casualties
in traffic. Lives can be saved by using seatbelts. You should never make
jokes about that." Van Nieuwenhuizen continued: "I was really pissed.
Wearing seatbelts in the back has been mandatory since 1992. Every
Dutch person has to comply with the law and a minster should set an
example." The minister for economic affairs humbly offered excuses and
paid an amount equal to a traffic fine into a trust for traffic safety.[23]

Apparently, the minister's own sense of ethics was insufficient to avert this
public disgrace. But worse, and illustrative for the lack of a discussion on
risk, was that none of the people surrounding the minister were able to
safeguard him from this humiliation. Neither his driver, the civil servants
around him, his fellow ministers, nor even his family called out the physical and reputational risks of not wearing a seat belt. In this example, it was
only after a journalist spilled the beans that the minister bettered his ways.
In your organisation, you don't want to rely on the press or whistle blowers
to signal scandals to you – you want to pre-empt disaster by having an open
eye for hazards and lively discussions on risks.[24]

5.3.2 Building Expertise

You know you are on the right side of the curve if operators share their pride
in workmanship with you, and you are able to celebrate expressions of expertise with them.

There are no definite criteria to characterise those with expertise, but the
following might give you some clues: successful past performance, respect by
the peer group, ability to explain phenomena clearly, consistency in predictions, equitable reflection on recent performance and the ability to detect and
recover from errors before they become disruptive. Expertise can be shown
in unconventional ways, such as a paper cup on the flap handle of a big jet to
remind the pilot it still needs to be set; the wire tie around the fence so the
train driver knows where to exactly stop to dump his load; beer handles on
otherwise identical controls in a nuclear power plant to differentiate between
them; and the use of printer paper on overhead lights to reduce glare on
computer screens. In these ways, people 'finish the design' so that errors are
avoided, and things go well rather than badly. Expertise is basically the ability to recognise patterns and invoke an appropriate action. Decision-making
by experts has been shown to be triggered by recalling previous experiences,
even if this occurs subconsciously. It is sometimes labelled 'intuition' but it
is hardly a mystical skill. Although expertise doesn't transfer to a completely
different context, elements of expertise will often still be valid in new situations. For instance, despite Covid-19 being a new virus, virologists were still
able to apply much of their generic skills under these circumstances.[25]

The development and retention of expertise is crucial in all industries.
Competence decreases over time, and more so in case of a shortfall of the

initial training, lack of practice, task complexity, inexperience and age. This illustrates the need to build experience and to mentor the less practised in a way that stretches their expertise, allowing them to absorb the unexpected at some time in the (hopefully distant) future. Take this example:[26]

> In December 2019 an Airbus A320 operated by Jetstar Airways approached Ballina/Byron Gateway Airport in Australia during a day with calm and clear weather. The crew decided to let the co-pilot conduct a visual approach to practice his manual flying skills. As they were executing this, the crew realised that they were descending more rapidly than was prescribed and so conducted a go-around. This was good practice, because crews don't often get that chance. The aircraft generated a few error messages because the go-around was executed with a little too much thrust, which distracted from the standard procedures. These distractions also led to the missing of the 'gear down' item on the checklist, for which the crew is usually warned anyway – except when they have extended it once already close to the ground as was the case here. The automation caught the lack of landing gear anyway (as it should), albeit a little later in the sequence, now necessitating a second go-around. A light aircraft had entered the circuit and the crew coordinated with it, but due to the close proximity around the runway the Traffic Collision Avoidance System (TCAS) triggered nonetheless. The crew was able to land without further complications, and (I guess) with some valuable lessons learned.[27]

The airline in this case seemed to be 'not amused' and required both crew members to attend 'debriefings' with flight operations management and specific simulator and airborne training after the occurrence. With this reaction, is the airline helping its pilots build expertise? Rather, it seems that management is curtailing the creation of the very capacity that allows the airline to sustain operations when things get difficult. It doesn't help of course that the press got wind of this event and labelled it as 'narrowly avoiding a major disaster'...[28]

Expertise is built by trying things out at the edge of the operating envelope in a safe-to-fail manner. As inventor Richard Browning of jet suit fame says: 'We learned by trying things and learning by failing at them most of the time. And that included failing by falling over. [...] And then I pinched a fuel line. So again, good learning. We learned not to do that again.'[29]

But then he made sure that he only fell a few feet at a time, that is failing (falling?) safely. Peer feedback and task-centred coaching are another way to maintain and advance expertise. Some senior professionals will not have had detailed, sincere feedback on the way they execute their tasks for years. Organising reviews where senior managers get to challenge the performance of each other's department or division is another way to maintain and advance expertise. Overcoming the reluctance of professionals and getting this organised will take a little perseverance but the rewards can be incredible. It is also possible to disseminate expertise so that less experienced staff can quickly climb on the learning curve. Rather than being instructed, it requires them to be confronted with novel situations, and to think through the options. Expert feedback is then used to calibrate their thinking.[30]

5.4 Conclusion

Drift (or the erosion of safety margins) is a natural occurring phenomenon that is the result of the perceived success of the organisation in the absence of significant incidents. Five factors make drift inevitable in any organisation: scarcity and competition, decrementalism, sensitivity to initial conditions, unruly technology and failing protective structures. The adaptations that are made by the operators to accommodate and correct for the increased pressures to achieve their goals create the illusion that the system is performing well and that the increased risks resulting from these demands are acceptable. The tell-tale signs of drift are window dressing for audits and inspections, experts knowing things that you do not, and only finding out about how work really gets done after adverse events. Drift can be countered by ensuring that discussions about risk are kept alive – through diversity, lack of hierarchical and departmental barriers, by deferring to expertise and getting outsider views. The result is improved safety, but that at times, operations are delayed until resources or controls are added or circumstances are more favourable. Finding the balance between these is – as always – the challenge, but you don't want to rely on journalists and whistle blowers to tell you you are setting your priorities incorrectly. Building expertise also helps to counter drift. This is achieved by practising at the edges of the envelope, facilitating sincere feedback and sharing experiences.

Key actions:
- Create diversity throughout your organisation.
- Ensure that expertise carries weight in discussions.
- Foster diversity of opinion, and facilitate vigorous debates even if it takes more time.
- Allow discussions to result in stopping operations, adding assurances or requiring further resources.
- Ensure that the discussions are founded in common mission and values.
- Revisit previous outcomes to identify whether the right balance was found between thoroughness and efficiency and learn from this.
- Request proactive risk assessments by safety professionals for new or modified circumstances.

- Require people to practise their skills and ensure that they are challenged to build and maintain their competences.
- Promote mentoring, peer feedback and task-centred coaching, even for the proficient.

And still things will occasionally go wrong. That is the subject of the next chapter.

Notes

1. ... balance efficiency and thoroughness...: Hollnagel (2009).
   ... mask what is really going on...: Patterson and Wears (2015).
   ... as in 'drifting into failure'...: This chapter builds on Sidney's renowned book: Dekker (2011).
2. ... by five factors...: Dekker (2011).
3. ... All organisations...: Jens Rasmussen (1997) modelled the limited space for businesses to manoeuvre in, bounded by the constraints of workload, economic survival and safety. He suggests 'Brownian-like' movements across this space as organisations seek to optimise their performance. Note that in a complex environment, it is impossible to know whether full optimisation has really been achieved. We can initiate local optimisation in one area (e.g. budget cuts in support departments) but this may lead to additional resources needed – or a concealed increase in risk – elsewhere. Hollnagel (2009) simplified Rasmussen's space to a balance between efficiency and thoroughness, showing through numerous examples how the scales slowly favour efficiency over safety until an incident jolts the balance back.
4. Amernic and Craig (2017).
5. Dutch Safety Board. (2019). Fire at Esso, 21st of August 2017 [in Dutch].
6. ... running aground of the Costa Concordia...: Anand (2020); Di Lieto (2015).
7. University in the UK, own experience, 2019.
8. ... Cases like this occur...: Patterson and Wears (2015).
9. Oil & gas industry, own experience, 2018. The company used this example as a trigger for a critical internal discussion between management and operators about minimum staffing levels versus workload.
10. ... accustomed to deviant behaviour...: When people within the organisation become so accustomed to deviant behaviour that they don't consider it as such, despite the fact that they exceed their own rules for safety, it has been termed 'Social Normalisation of Behaviour'. Vaughan (1999).
11. Dutch Safety Board. (2015). Explosions MSPO2 Shell Moerdijk, June 3rd, 2014.
12. ... sensitivity of safety to initial conditions...: Leveson (2015, p. 24) suggests that there are six domains from which safety-critical assumptions generally arise:
    - The hazardous states that have (not) been identified for the system.
    - The effectiveness of actions to reduce or manage hazards.
    - The way the system will be operated and the environment in which this happens.

- The way the system has been designed and developed.
- The organisational control structure during operations, including the safety culture and whether policies are being followed.
- The way that risks are managed, including how these leading indicators are used.

It is important to ask for each of the domains: How does the current state of the system deviate from what was originally intended? Leveson suggests that 'leading indicators' be devised that signal whether initial assumptions still hold for the system of interest. She gives as an example (p. 32) the need for independent checking of design deliverables. But when there is a significant engineering backlog, these checks may be bypassed, so that one indicator for the validity of the original assumption is the amount of work-in-progress in the checking process for engineers.

13. ... does not match the operator's sense...: This is referred to as an operator having a buggy mental model, and can lead to an Automation Surprise: the conscious and sudden realisation that what is being observed does not fit the current frame of thinking. This occurs after relevant cues from the environment have been ignored due to an existing frame or mental model, and then a sudden awareness occurs of the mismatch between what is observed and what is expected. In a sensemaking model that helps to explain this process, *surprise* marks the cognitive realisation that what is observed does not fit the current frame of thinking. Our own research suggests that in these cases, we need to take systemic factors and the complexity of the operational context into account, rather than focusing on suboptimal human performance. The gap between system state and our realisation of it seems to be a manifestation of system complexity and design choices and rarely the result of either individual underperformance or a failure in the automation.

    de Boer, Heems and Hurts (2014); Rankin, Woltjer and Field (2016); de Boer and Dekker (2017).
14. Dutch Safety Board (2010); Dekker (2009).
15. ... 'Automation Surprises' occur...: Woods and Hollnagel (2006).

    ... once every three months...: de Boer and Hurts (2017).
16. ... protective structures...: This includes anything that is aimed at maintaining safety margins, such as assurance processes (regulatory arrangements, quality review boards, safety departments), assurance information (reporting systems, meeting notes, databases) and staffing choices (previous experience, affiliations, ambitions, renumeration and social bonds). Dekker (2011).
17. Gelles, D., & Kitroeff, N. (2019). Boeing hearing puts heat on F.A.A. Chief Over Max Crisis. *The New York Times*, December 11th, 2019. https://www.nytimes.com/2019/12/11/business/boeing-faa-737-max.html;

    Gelles, D. (2020); 'I Honestly Don't Trust Many People at Boeing': A broken culture exposed. *The New York Times*, January 10th, 2020. https://www.nytimes.com/2020/01/10/business/boeing-737-employees-messages.html.
18. Oil & gas industry, own experience, 2018. The company has since improved its audit regime to enable more learning using micro-experiments. See paragraph 4.3.

    ... the prescribed isolation of process equipment...: Isolation is used to separate machinery and equipment from an energy source or (dangerous) substances such as oil or gas. It means that the machinery is physically or through valves or controls

disconnected from its surroundings and used to allow repair, service or maintenance work to the equipment without endangering operators or the environment.
Own notes, October 2018.

19. ... Remember also...: Internal email, English university, January 2020.
20. ... sustaining their operations...: Resilience is 'the intrinsic ability of a system to adjust its functioning prior to, during, or following changes and disturbances, so that it can sustain required operations under both expected
and unexpected conditions' (Hollnagel 2013, p. xxxvi). Hollnagel suggests
that resilience relies on the capacity of individuals, teams and organisations
to adapt to a threat or a changing work environment and draws attention to
the ingenuity and adaptability that professionals display to maintain ordinary and apparently 'standard' operations under challenging and variable
conditions. In the following paragraph, I suggest how this ability can be
developed.

   ... the edges of the operating envelope...: A term to describe that operations
are only just within prescribed limits, regarding the weather, available resources,
process conditions, etc. This is where safety margins are considered to be the
smallest. Dekker (2011, p. 178).

21. ... several characteristics...: Adapted from Dekker, S. (2018). Why do things go
right? https://safetydifferently.com/why-do-things-go-right/, accessed January
22nd, 2020.
22. ... revisit the outcomes of earlier discussions...: Time and again we have been
confronted with employees whose safety concerns have been overruled by
senior managers – usually without adverse consequences. However, we rarely
hear that afterwards the event is discussed, to determine whether staff were
too cautious, or the situation was fortuitous. This revisiting of the situation is
necessary to ensure that managers continue to listen and engage with experts
around them, and remain sensitive to risks.
23. Van Zwienen, S., & Winterman, P. (2019). Gordelweigeraar Wiebes gaat door
het stof en maakt boetebedrag over aan Veilig Verkeer Nederland. Algemeen
Dagblad, September 27th, 2019. https://www.ad.nl/politiek/gordelweigeraar-
wiebes-gaat-door-het-stof-en-maakt-boetebedrag-over-aan-veilig-verkeer-
nederland~a235b133, accessed March 2nd, 2020.

   More recently, the UK advisor to the Prime Minister was caught up in a similar example, where he broke the stay-at-home rules to constrain the corona
virus and was caught by the press. https://en.wikipedia.org/wiki/Dominic_
Cummings#COVID-19_pandemic, accessed June 22nd, 2020.

24. ... you don't want to rely on...: Actively seeking out criticasters is one way
to guard against drift. Dekker (2011, p. 175) even goes so far as to suggest a
measure that can be used to identify shortfalls in diversity called 'Herfindahl
index'. One need only think of the discussions preceding the accident with
the space shuttle Challenger, where the voice of concerned engineers was
not heard by those taking the decision to launch. Rogers Commission, (1986,
p. 83).

25. ... no definite criteria to characterise those with expertise...: Expertise is similar to skill but emphasises cognition. Klein (2018) mentions seven criteria:

   1. 'Successful performance – measurable track record of making good
   decisions in the past. (But with a large sample, some we do very well

just by luck, such as stock-pickers who have called "the market direction" accurately in the past 10 years.)

2. Peer respect. (But peer ratings can be contaminated by a person's confident bearing or fluent articulation of reasons for choices.)

3. Career – number of years performing the task. (But some 10-year veterans have 1 year of experience repeated 10 times and, even worse, some vocations do not provide any opportunity for meaningful feedback.)

4. Quality of tacit knowledge such as mental models. (But some experts may be less articulate because tacit knowledge is by definition hard to articulate.)

5. Reliability. (Reliability is necessary but not sufficient. A watch that is consistently 1 hour slow will be highly reliable but completely inaccurate.)

6. Credentials – licensing or certification of achieving professional standards. (But credentials just signify a minimal level of competence, not the achievement of expertise.)

7. Reflection. When I ask "What was the last mistake you made?" most credible experts immediately describe a recent blunder that has been eating at them. In contrast, journeymen posing as experts typically say they can't think of any; they seem sincere but, of course, they may be faking. And some actual experts, upon being asked about recent mistakes, may for all kinds of reasons choose not to share any of these, even ones they have been ruminating about. So this criterion of reflection and candor is not any more foolproof than the others'.

However, 1. and 3. seem similar and 6. appears a threshold (and incorporated in 2.) rather than a criterion. I prefer to reword 'quality of tacit knowledge' as ability to explain phenomena clearly, as the former is hard to judge. I assume that Klein means 'consistently successful in predictions' when he mentions 'reliability'.

Klein, G. (2018). How can we identify the Experts. Seven criteria for deciding who is really credible. Blog for Psychology Today, available at https://www.psychologytoday.com/us/blog/seeing-what-others-dont/201809/how-can-we-identify-the-experts, accessed June 27th, 2020.

... the ability to detect and recover errors...: Allwood (1984).

... paper cup...: Dekker (2014b, p. 158).

... the wire tie...: Dekker (2019, p. 409).

... beer handles...: Dekker (2019, p. 409).

... to reduce glare...: Own experience, Hogeschool van Amsterdam Wibauthuis building, August 2019.

... Expertise is basically...: Schraagen, J. M. (2020). Who are the 'experts' in Covid-19? Webinar for the Resilience Engineering Organisation, May 11th, 2020, available at https://www.resilience-engineering-association.org/seminars/, accessed June 27th, 2020.

... Decision-making by experts...: Klein, Orasanu, Calderwood and Zsambok (1993).

... hardly a mystical skill...: Schraagen (ibid).

26. ... Competence decreases over time...: Vlasblom, Pennings, van der Pal and Oprins (2020).

... absorb the unexpected...: Not surprisingly, pilots have more difficulty in managing an emergency when this situation is presented unexpectedly. Landman, Groen, Van Paassen, Bronkhorst and Mulder (2017).

27. Adapted from Carim Jr, G. (2020). You See Pilot Error, I See Normal Work. https://safetydifferently.com/you-see-pilot-error-i-see-normal-work, accessed February 16th, 2020.

28. ... The airline...: Australian Transport Safety Board (2019).
    ... 'narrowly avoiding a major disaster'...: Mazzoni, A. (2019). Jetstar pilots forgot to lower the WHEELS as they came into land – narrowly avoiding a major disaster. *Daily Mail Australia*, December 11th, 2019. https://www.dailymail.co.uk/news/article-7779275/Jetstar-pilots-forgot-lower-wheels-came-land-Ballina-Byron-Gateway-airport.html, accessed February 22nd, 2020.

29. ... As inventor Richard Browning...: Browning, R. (2017). How I built a jet suit. Ted talk, Vancouver, April 2017. https://www.ted.com/talks/richard_browning_how_i_built_a_jet_suit, accessed April 23rd, 2020.

30. ... Peer feedback and task-centred coaching...: Gawande, A. (2011). Personal Best Top athletes and singers have coaches. Should you? *The New Yorker*, October 3rd, 2011. https://www.newyorker.com/magazine/2011/10/03/personal-best, accessed April 13th, 2020.
    ... to disseminate expertise...: Klein, G. (2020). Naturalistic Decision Making Perspectives on Covid-19. Webinar for the Resilience Engineering Organisation, May 11th, 2020, available at https://www.resilience-engineering-association.org/seminars/, accessed June 27th, 2020.

6

What to Do When Things Go Wrong

6.1 Introduction

Adverse events happen, perhaps because of the introduction of new technologies, changes in the way we work, societal developments or economic pressures. In this chapter, we discuss how we can ensure that we are cognisant of incidents that occur despite all our efforts to maintain safety, how we can maximise our learning from them and how to restore the relations that suffer as a result of adverse events.[1]

Even if the gap between *paper* and *practice* is managed to a minimal level and we apply our best efforts to counter eroding safety margins, still occasionally things will go wrong. Although it is straightforward and therefore attractive to focus on personal behaviour as the most important source of safety, this ignores much of the benefit that can be achieved by delving deeper and widening the scope. There will always be at least some shared culpability through the design of the task, the way that work is supervised and the conflicting goals that we expect our employees to meet.

> On February 10th, 2010, a Boeing 737 took off by mistake from the taxiway parallel to the assigned runway at Amsterdam's Schiphol airport. Due to a variety of confusing cues the crew mistakenly believed they were at the runway threshold. There was fortunately no other traffic on the taxiway as they started their take-off run, and they departed for Warsaw without further problems. Although press releases seemed to focus on the culpability of the crew members, in fact the investigation report notes only one personal factor that contributed to the incident (not using the ground movement chart) and seven infrastructural and policy factors: unusual taxi route, late taxi route changes, need to reprogram the flight computers during taxi, abnormal radio communication, no monitoring of taxi track by air traffic control, and taxiway lighting and markings that were difficult to distinguish in the prevailing conditions.[2]

Note that the infrastructural and policy factors mentioned in the example above can (often exclusively) be addressed days and weeks ahead of the incident, by people and organisations at a distance to the flight crew. These external entities probably did not realise that they had set the crew up for failure until after the incident occurred. Often, we rely on those who are directly involved in a task to save the day, even though much of error prevention and safety promotion is dislocated in time and space from an incident. As management, we do not usually hear about the error traps that we set for our employees until very late. Some error traps might be due to hasty task design and limited testing, but many others are due to fickle interactions that can only be experienced in real life. Without us realising, there are error traps everywhere. More often than not, and probably without us knowing, we ask our employees to 'finish the design': to adapt the rules and tools that we equip them with so that these function properly in practice, and error traps are avoided.[3]

6.2 Hearing about Incidents

In the second chapter of this book, I have addressed the need to welcome 'bad news' as a chance to learn, not as something that damages your reputation. You therefore need to have a reporting system that allows your people to signal where work is difficult or dangerous. Reporting systems rely on the willingness of front-line employees to report a gap – one they might have no interest in sharing or not even perceive as dangerous, so we need to make it as easy and attractive as possible for them to submit a report.[4]

In highly regulated industries, there will be a formal reporting system supported by technology and perhaps dedicated personnel to make the system work. In other industries, it may be a mailbox fixed to the wall in a hallway or canteen, or the CEO soliciting improvement suggestions through a special e-mail address. In each case, the same factors are important that facilitate reporting and learning from incidents: foremost a clear vision that reporting is important for the organisation, and (only) secondly an appropriate reporting system to facilitate this.

The reporting system is one of the best means by which to hear about gaps between *paper* and *practice* and to capture bad news, and – independent of its formality – should be:

- Voluntary;
- Non-punitive;
- Protected (confidential);
- User-friendly;
- Accessible (system close to workstation);
- Generating timely feedback to reporter.

The voluntary nature is important because you want to be sure that reports are not submitted just to make the target – we have seen this happen when a goal has been set for a minimum number of reports a month. Non-punitive means that the report or any investigation following from it cannot lead to punitive action of any kind – even unintended or indirectly. If you do not abide by this suggestion, then naturally you will drive much of what happens underground. The reports need to be confidential, so that only a limited group of safety professionals, not line managers, know from whom the report derives. The reports cannot be anonymous because that countermands any opportunity for further investigation, and it invites malicious reports. Of course, the system needs to be really user-friendly and accessible, because many reports need to be made on the fly.[5]

> At one airline where we were asked to report on safety performance, we found that that the formal reporting system was not used sufficiently. The employees did not believe that a formal report would be treated confidentially, were afraid to offend others and thought that they would not be taken seriously. Instead, a local messaging service had taken over much of this functionality. Several chat groups were used by employees and managers to inform each other. As a manager stated: "The [chat] group is more informative than the reporting system. Anyone can express their opinion right away. [It is] faster than the reporting system." Of course, the messenger service was not integrated into the safety management system and so it does not support the exploitation and aggregation of the information. Therefore, opportunities were missed at this organisation to make safety management more proactive.[6]

Finally, any reporting system will be untenable if the reporters do not get the sense that their reports are useful – even if a report is not investigated, this needs to be fed back to the submitter.

Overall, the reporting policy – and management itself – needs to emit the conviction that bad news is welcomed, in fact even craved for. This is less difficult to put on paper than into practice but is *the* key to pre-empting incidents. Professionals tend to prefer to fix and forget incidents that they can resolve themselves. Near-misses may be considered unworthy of reporting since they do not result in actual harm. Many incidents are considered unique so that their value for learning seems limited. Re-occurring safety problems may seem as inevitable routine events. It is therefore quite a challenge to have as many incidents reported as possible, so that we maximise our collective intelligence about risks.[7]

6.3 Understanding the Event

Following an adverse event, we need to establish how work was done and comprehend why it made sense for the task executors at the time to act as they did. Any judgement needs to be upheld, and exemptions from the procedures need to be – at least initially – regarded as shortfalls of those procedures rather than as a 'violation'. This analysis resembles that of a gap between *paper* and *practice* but is much more difficult because emotions as guilt, blame and remorse usually distort the picture after an adverse event. There is the pain of injuries or worse, and apprehension for liabilities and costs for damages. Severe incidents tend to generate a lot of attention, also from groups possibly not familiar with emergent behaviour in complex environments and the difference between *practice* and *paper*. Higher management, authorities, press and general public might consider *all* incidents and accidents to be avoidable and require that someone is blamed, which is hardly helpful in learning and improving.[8]

The objective of the first phase of the investigation is to understand and document *what* happened, *how* it happened and *why* it happened. I will also discuss when fact-finding can be deemed to be complete, and the language that might be used in the report.[9]

To understand *what* happened, it is necessary to collect evidence, identify the facts and define a chronology of events. Data sources include photographs, video material and sound or voice recordings, interviews of witnesses, victims and operators of the system, documents, recordings of the system state, geographical information and an analysis of the consequences of the event and the damage. It is often powerful (if relevant) to create a chronology, a site map and a schematic of the hardware. Pitfalls in fact-finding include the following:[10]

- Preconceived notions of what happened and adherence to initial observations, rather than extending the scope by considering alternatives;
- Contextual confusion of the situation, event or system state, possibly originating from groupthink or lack of peer challenge;
- Lack of resources that limit data collection and analysis;
- Deficient assessments due to a lack of knowledge;
- Lack of independence to challenge the existing processes and structures.

Naturally, the value of the investigation as a tool to aid learning will be compromised by the above. It is often useful to refrain from pre-empting causes until the investigation has been completed because only then can the full story be told.

In January 2019, a small aircraft was flying from Nantes, France, to Cardiff, Wales when it crashed in the English Channel off Alderney, Channel Islands. It was carrying the Argentine football player Emiliano Sala who had just transferred to Cardiff City football club for €18 million and a pilot friend. The wreckage of the aircraft was found a few weeks later, and the investigation team released an interim report that mentioned that neither the aircraft nor the pilot were certified for a commercial flight, and that the contribution of the weather conditions in the area at the time of the accident were subject to further examination. In August 2019 it was reported after post-mortem tests on Sala's body that exposure to carbon monoxide was the most likely cause of the accident.[11]

As can be seen from this example, the probable causes of the accident shifted dramatically after the autopsy of Emiliano Sala. Drawing premature conclusions after the intermediate report would have led to incomplete and possibly unjustified recommendations.

To answer the question about *how* the event happened, it is often necessary to assume a certain causality between facts and to assess the plausibility of different scenarios. This step can include the use of methods that support the analysis by pointing to possible weaknesses in the wider context of the event that may not have been immediately evident. These methods include an analysis of organisational factors, the role of regulatory bodies or even those of government, and the effects of variability on normal operations. The results of an analysis utilising one or more of these methods are therefore richer than if these methods are not utilised.[12]

The word 'finding' is often used in the context of incident investigations. A finding is the discovery and establishment of the facts of an issue. Findings can be facts that have been verified, presumptions that have been proven to be true or false (based on the available facts or analysis) or judgements beyond reasonable doubt when dealing with human and organisational factors. It is usual to formulate hypotheses – possible findings that need substantiation – immediately after fact-finding. They help us to prioritise and/or reduce the amount of work. Hypotheses that are unlikely to be true are just as important as hypotheses that are likely to be correct. Hypotheses that are substantiated and are deemed relevant become findings. The findings should highlight the major factors that contributed to the event sequence and are often grouped into conclusions.[13]

To understand *why* an event occurred, it is quite often necessary to recognise the rationale for an actor's actions and decisions within the context of the event. As described earlier, there is little value in suggesting 'an error' as the cause of the adverse event. *Why* did it make sense for the actors to do what they did? Don't forget that nobody comes to work to do a bad job (if they do, then it is outside the scope of the safety domain), and they didn't know the outcome of the event while they were executing their task. There will be a local, bounded rationale that explains the actions of those involved, based on the goals, resources and constraints that were available to them at that time. Note that it is sometimes difficult to get into someone's head retrospectively, and therefore, there is always an element of speculation in determining *why* courses of action made sense to the actors.[14]

Together the findings define *what* happened, *how* it happened and *why*. Stakeholders should not be involved in identifying the findings and deciding upon the most plausible scenario, although they can be asked to check for factual errors. They will be biased towards maintaining the status quo and suggesting that the incident was an abnormality. The established findings should provide a robust foundation for the organisation to learn from, and to identify and formulate recommendations.

When to stop an investigation and consider it complete is a professional judgement call: there is no 'correct' or absolute set of causes for any given incident. Most investigators take justified pride in their work and would like to present a masterpiece. It is easy to get carried away and try to answer 'the ultimate question of life, the universe, and everything'. However, the purpose of an incident investigation is to distil lessons from the event that might help prevent reoccurrence, and so a stopping rule is required that enables learning yet minimises costs. The development needs of the organisation rather than personal interests should guide the investigators in deciding when enough is enough. The discovery, establishment and scope of findings to be included in the final report are choices that the investigator makes – either consciously or unconsciously. There are no definite truths about what 'caused' an event (and certainly not a single 'root' cause). Rather, the investigator constructs a most plausible scenario that is useful in helping the organisation to understand what happened and to learn from the event. In any complex environment, it will include multiple contributing factors, including several that are dislocated in time and location from the event itself. Because the choices that the investigator makes are so important, these should ideally be reviewed by a group of peers and justified in the report. The belief that the investigators are working to support the company in improving its performance (rather than flaunting their expertise) is crucial to securing the acceptability of the findings.[15]

The language with which the investigation report is written reflects the attitude of the investigating team. I encourage you to consider the following when composing the report to maximise concurrence for it and its learning potential:[16]

- Describe **'human error' as a symptom** of something wrong in the system, rather than a cause for the adverse event. Describe why it made sense for those involved to act as they did and focus on what made them think this was a normal or the best course of action. Poor example: 'Despite indications in the cockpit, the cockpit crew did not notice the too big decrease in airspeed until the approach to stall warning.' This sentence begs the question *why* the cockpit crew did not notice the decrease in airspeed.
- **Avoid retrospective biases** such as hindsight bias and outcome bias. These are closely related – hindsight bias overestimates one's ability to have predicted an outcome; outcome bias unfairly judges a decision on information only available after the decision's outcome is known. Poor example: 'In the situation that had arisen, the parties

involved failed to recognise adequately the risks involved in offering and accepting [the change], as a result of which the procedures were not carried out with the necessary close attention required in this particular case [emphasis added]. That this case required closer attention than usual was only known in retrospect.'

- **Avoid counterfactual descriptions**: things that did not occur but could have occurred, for example 'The crew did not monitor the aircraft's position using a ground movement chart'. Counterfactuals are often a result of too much faith in Work-as-Imagined and an indication that the gap with Work-as-Done has not been sufficiently investigated. Counterfactual descriptions do not help us understand why the adverse event occurred (in this case, a take-off from a taxiway instead of the runway). We need to understand how this aircraft crew usually navigates around their home airport and whether their way of doing it is common practice.

- **Avoid passing judgement** in the report. The intention of the report is to identify areas of learning and improvement, not to apportion blame. Passing judgement is not only unfair (an example of retrospective bias) but also very counterproductive for this and any future investigation, because it triggers defensiveness which is a barrier to learning. Poor example: 'In this case, the drew did make such an error, which the runway controller subsequently failed to notice on [sic] time'.

- Ensure that **distal** actors and events are taken into account, not only those decisions, actions and events that are geographically or in time close (proximal) to the adverse outcome. The contribution of these may be less immediately visible yet equally material. A good example is the mentioning of the need to improve signage at the airport to avoid aircraft pilots inadvertently taking off from a taxiway.

- Compare the actions leading to the adverse event with **routine actions** that did not lead to an adverse outcome – these may be very similar and are therefore not a differentiating cause of the event. In this case, it turned out that pilots 'usually do not use [a ground movement chart] at [their home airport] because they know the airfield well and use their common sense to determine what is and is not necessary'. The non-use of the ground movement chart is therefore not a differentiating factor between this incident and routine work.

- Avoid simplifying the context surrounding the adverse event, and ensure that the **complexity of the system** within which the adverse event arose is sufficiently described. For example, was emergent behaviour recognised and were shortcomings in the interactions between actors identified before it could lead to an adverse outcome (rather than focussing on individual errors)? In this example it was appropriately noted that, the air traffic controller was forced to shift his attention to another aircraft 'because the runway controller suddenly had to solve a problem', even though it is prescribed that an aircraft'sposition is monitored continuously while on the ground. On a more abstract level, the report fittingly mentions goal conflicts: 'the incident was caused by the decision to follow a shorter route [...] aimed at stimulating the flow of traffic. Punctuality was also important for the pilots.'

- Do not use wording that mystifies rather than clarifies (sometimes called **'folk models'**). An example is: 'While taxiing the crew lost their *positional awareness* as a result of which they took off from taxiway Bravo instead of the adjacent take-off runway' [emphasis added]. The use of this folk model suggests a cause but in reality it does not help us understand the crew's action – we now need to ask why they lost their 'positional awareness'. Either replace the folk model with an actual cause (if known) or eliminate it altogether to maximise information density. For example, 'while taxiing the crew took two right turns as they always do, as a result of which they took off from taxiway Bravo instead of the adjacent take-off runway' or 'The crew took off from taxiway Bravo instead of the adjacent take-off runway without realising this'.[17]

Abiding by these suggestions will significantly contribute to the objectivity and readability of the investigation report and therefore its learning potential.

6.4 Maximising Learning

The purpose of an incident investigation is to distil lessons that might help prevent reoccurrence of similar events. That means that the recommendations that follow from an investigation need to be tailored to the organisation and their development needs: there is no 'correct' or absolute set of recommendations that follow from any given incident. Identifying how best to maximise learning is a collaborative responsibility of the investigators and those being addressed by the recommendations. In this section, I expand on the way that this might be successfully accomplished.[18]

Recommendations follow directly from the analysis and findings. They suggest applicable corrective actions that are aimed at preventing similar events from happening again here and elsewhere; that mitigate the consequences should such an event happen again in future; that address knowledge deficiencies revealed during the investigation; and that tackle weaknesses in the relevant human, technical or managerial processes that led up to the event. Recommendations need to be based on the conclusions but not all conclusions require a recommendation to follow. Making effective recommendations requires appropriate knowledge of the wider system. Towards this aim, and to maximise buy-in, it is useful to consult stakeholders early while developing recommendations. Involving them leads to more practical recommendations and therefore a higher probability of them being implemented. Actually, some suggest that 'the really big safety payoff occurs when everyone agrees during the course of an investigation about what needs to be done. In these situations, safety deficiencies are quickly addressed. All we need do is to document in our final report the actions already taken'.[19]

The investigator needs to realise that the most valuable and difficult part of an investigation is not the investigation itself, but the mitigation of the identified risks. Recommended actions can vary in effectiveness: elimination or physical isolation of hazards is generally more effective than warning signs or changes to rules. However, these are not always feasible and generally more costly, so that recommendations for administrative improvements (including training) are four times more common – despite the fact that these improvements are much less likely to be effective in preventing recurrence. It's probably also good to realise that some 'improvements' actually complicate operations and might even increase the likelihood of new types of accidents in future.[20]

It helps if recommendations are clear and unambiguous even when they are read separately from the rest of the report. The investigation report should clearly explain how the recommendations were reached and how they are supported by the evidence of what happened. Recommendations ideally are not prescriptive because it is line management, not the investigator, who must work out how best to eliminate, or significantly reduce, the safety deficiency. It is good practice to include a realistic time limit for responding to a recommendation so that progress can be tracked. It is the organisation's responsibility (i.e. its higher management), not the investigator's, to hold people accountable for the implementation of improvements and the mitigation of risks. Monitoring progress can definitely help to keep the discussion on risk alive, but in the end it is the responsibility of line management to determine whether and how these are managed and at what costs. Effective investigators are those that understand that the choices that the investigation team makes in identifying the causal factors are subservient to the learnings that can be achieved in a specific context, and can judge where improvements are most challenging yet feasible for the organisation.

Having gone through the pain of an adverse event and paid the costs associated with the investigation, it seems like a shame not to reap the full fruits of this. It is therefore highly recommended to review the implementation of the recommendations, but *particularly their effectiveness*. We often encounter a bureaucratic devotion to the risk log and whether actions have been executed, but hardly ever a revisit as to the effectiveness of the entries. The consequence is a long list of overdue actions that are low on priority because there is not really a belief in their usefulness. We would rather see feedback on the effectiveness of measures, a continuous debate about risk, a purge of unnecessary actions – and an abolishment of meetings in the name of safety that are just 'going through the motions'.[21]

6.5 Restoring Relations

In Chapter 2, I introduced an approach to adverse events that avoids retribution. This approach (called 'Restorative Practice') is aimed at restoring relations that have been damaged in the wake of an adverse event. It addresses the needs of direct victims but also of practitioners who feel responsible, and of the wider organisation. Besides catering for direct victims, it relies on the recognition by the practitioners of the physical and reputational damage that is done to the organisation, as well as management's acknowledgement of the pain and remorse experienced by these practitioners. This process takes time and is effortful, but has been shown to reduce blame, enable transparency, improve motivation and create economic benefits. The advantages of restored relations for learning are evident: the experience surrounding the event is retained for the organisation and often those involved will help to disseminate learnings. There are multiple examples where the direct victims of an adverse event prefer to be heard rather than claim for financial compensation. However, restoring relations after a major incident is not easy, and several constraints may stand in the way. I will first present the Restorative Practice methodology and then discuss the conditions to make it possible.[22]

Restorative Practice differs from a traditional 'Just Culture' in many ways. Where a 'Just Culture' has been premeditated by management and subjugation is not at the discretion of individual employees, participating in a Restorative Practice approach is voluntary and in fact requires deep commitment. Restorative Practice aims to restore relations, and we acknowledge and involve a much larger group compared to organisations applying a traditional 'Just Culture'. We strive for full transparency. A 'Just Culture' utilises algorithms, whereas a restorative approach involves a dialogue. Restorative Practice asks who is hurt and what their needs are. It looks forward by assessing who can, or should, meet those needs. It invites all affected to tell their accounts of the harm and their needs. To do so, it invests in relationships between people whose work depends on each other and repairs trust. It learns and prevents by asking why it made sense for people to do what they did. It meets hurt with healing.[23]

The difference between a retributive paradigm (paying compensation) and Restorative Practice (being heard) is evident in the case of Adrienne Cullen:

> Adrienne Cullen died of cervical cancer in 2019 after the result of a biopsy in 2011 (which tested positive) had not been returned to her gynaecologist until 2013 (!). At that stage her cancer had spread making recovery impossible. The board of directors of the hospital in Utrecht, the Netherlands, refused to talk to ms. Cullen, instead relying on lawyers to communicate, and did not investigate the error. In 2015 the hospital offered compensation to her and her husband. "When I finally got to talk to the board of directors, I said they were driving cars that cost [that much]. Was that all my life was worth?" After press coverage of the case the hospital suddenly offered a lot more money. But to the surprise and chagrin of Cullen, she had to sign for secrecy, which she refused (she called it the 'gagging clause'). She wanted the world to know about her case and learn from it. Finally, in 2018, after years of perseverance (and just before the cancer ultimately took her life), ms. Cullen and her doctors spoke openly about the aftermath of the medical error. The hospital finally investigated the error and made the results public, acknowledged its remorse and accountability and promised to be more transparent and patient-centric in future. An annual lecture bearing her name was established and she was able to attend the first edition. [24]

The restorative approach is aimed at engagement of all involved parties, emotional healing, reintegration of practitioners and ultimately organisational learning. It may take some time to achieve all of these outcomes. To implement a restorative approach, it may be necessary to review and revise the policies regarding human resources, although mostly the changes are relatively insignificant. However, the 'Just Culture' flow chart that aims to select a fitting punishment must be binned, and a real effort needs to be made to ensure that managers understand and are aligned with the principles of the approach.

6.5.1 The Triggering Event

A restorative approach is invoked by an adverse event: an event that could have caused or did result in harm to people, including psychological harm, feelings of hurt, injury or death, damage to property or reputational harm. It is evident that a restorative approach is necessary in case of injury or death or significant damage to property or reputation. An event without physical damage is sufficiently adverse to trigger a restorative approach if emotions like anger, pain or remorse flare up and relations are damaged beyond immediate repair. You or your colleagues may only hear about an adverse event sometime after it has occurred. In that case, the restorative approach is still pertinent, to be invoked as soon as possible. [25]

After an adverse event, the situation needs to be stabilised first to ensure that no (further) physical or psychological damage can be done. This may require separating the direct victims and professionals that were involved in the event. It is often helpful to have someone coordinate the restorative approach, and schedule conversations with the key stakeholders as soon as possible – within hours rather than days. A kneejerk reaction in some industries is to respond immediately in the direction of the practitioners involved in the event by suspending or investigating them. The devastating effects of an investigation into the acts of an individual cannot be underestimated. It only seems justified to investigate the need for possible disciplinary action against an individual if you can decisively conclude that *all* of the below are valid:

- Specific rules were violated;
- These rules are workable;
- These rules were knowingly departed from;
- Peers do not deviate from these rules in similar circumstances;
- Appropriate training was available to the individual;
- The individual had been provided with supervision by senior managers that should normally be provided.

Based on these requirements, it seems that an investigation into an individual is therefore only seldomly justified.

6.5.2 Three Simple Questions

To apply a restorative approach, the first question to ask is: **who is hurt?** This will likely include:

- Direct victims suffering the immediate consequences of the event without having played an active role in it. Examples include patients, passengers, colleagues or the surrounding community of a process plant.
- Practitioners involved in the event who feel personally responsible and remorseful, and suffer as a result of this.
- Local management that feels hurt and let down by the practitioners and who may feel that their good standing within the organisation has been harmed.
- The organisation at large that may have suffered damages, but also experience reputational harm, for instance, towards the regulator, politics or the local community.

It is important to acknowledge the hurt that has been inflicted on each of these parties. This creates room for a dialogue and is therefore the first step in restoring relations. In particular, management needs to acknowledge that it is not an innocent bystander but is responsible for the circumstances surrounding the event. Management will probably have been pained by the incident. Note that some time may elapse before all victims are able to engage in this process, particularly those that have been severely affected by the event.

The second question is: **what are their needs?** Direct victims might want to know how the event could have occurred and may need some sort of compensation for the damages sustained. They will want to be reassured that amends are being made to avoid reoccurrence. Professionals involved in the event will generally want to have an opportunity to tell their version of what happened – and to explain why it made sense to them to do what they did at the time. They might need psychological support and compassion, protection from the public, the press and prosecution, and will generally desire some sort of security regarding their livelihood. Local management will need recognition for their feelings of disappointment before they are able to restore relations. Local management and the rest of the organisation need to be reassured that the adverse event will not reoccur, not only for themselves but also for them to convey to their stakeholders.

The third question is: **whose obligation is it to meet these needs?** The obligation of the direct victims is to participate in the restorative process and respect the professionals that were involved in the event. The practitioners are required to honestly and fully disclose what happened, give suggestions what to do to prevent recurrence, and play an active role in sharing these lessons. They will have to show remorse for what happened and express their feeling of responsibility. Local management and the rest of the organisation must allow these victims to tell their account of what happened, offer support and ensure that fixes are implemented. Management should protect the practitioners from outside agencies, which includes providing legal support if required. The organisation must not fire or sanction people for Restorative Practice to work, because that will severely harm relations and impede cooperation in future cases. Management should ask *what* is responsible for the incident, not *who*. A learning review should be executed that shows why it made sense for people to do what they did. Pathways to prevention need to be identified in collaboration with the victims and professionals. Management needs to ensure that these pathways to prevent reoccurrence are initiated and maintained, and to report the progress of implementation to the victims of the event and other stakeholders.[26]

After the three questions have been satisfactorily addressed and the needs of the different parties have adequately been met, the result of this approach is hopefully that relations are restored and that learnings are (being) disseminated. This may take any amount of time varying from days to multiple years, as hurt may emerge over time or resurface repeatedly. Restorative Practice only works if the needs of all those hurt by the adverse event are properly identified and if the obligations to fulfil those needs are met. The process will fail if the practitioners involved in the adverse event are not honest or do not show remorse, or if local management is not able to allow the professionals to tell their account or cannot implement the necessary solutions.

6.5.3 The Demanding Nature of Restorative Practice

The need to be honest and remorseful in the face of superiors and victims, and helping to disseminate the learnings of the event to colleagues, can be quite daunting for many, and should counter the misunderstanding that this approach is too lenient towards practitioners involved in an event. The injured parties need to (be made to) understand that such a confession is risky, and recognises it to be meaningful, respectful and worthy of a response. It may be agreed that the professional requires targeted coaching or training to avert reoccurrence, and performance monitoring may be intensified to ensure the interventions are effective.[27]

A restorative approach is also challenging for the supervisor of the professionals that are involved in an event. In a traditional 'Just Culture', a line manager can distance himself from the event and the consequences. In case of a retribution following an event, this can be easily justified and will be anticipated, even welcomed, by management. In contrast, a restorative approach implicates supervisors and requires their full participation, its outcome is uncertain, and (worst of all) emotions cannot be avoided.

The restorative approach is not an excuse to tolerate poor performance. Over time, it may become apparent that someone is not executing his or her tasks satisfactorily, even though different measures of support have been instigated. If termination in the current role is being considered, then the justification for this must become apparent during normal work, and definitely not from an adverse outcome. The latter is fraught with emotions and is at odds with the need to restore relations after an event.

6.5.4 If Restorative Practice Fails

It is possible that practitioners involved in an event are not able to uphold their obligations in the restorative approach. They may not be able to tell the truth or show remorse. In those cases, we see that organisations unfortunately often revert to the more traditional retributive approach.

Sometimes specific first victims (or practitioners that were subject to harsh treatment by the organisation) are not able to forgive and keep coming back on how they have been hurt, and it may not be possible to apply the restorative process. In that case, it may be necessary to consider an individual therapeutic trajectory, to protect the organisation against the negativity of residual hurt and to promote the support and enthusiasm for a restorative approach despite what happened. Eventually, it may be necessary to decide that these people are perhaps better off in another organisation and help them to acknowledge and realise this.[28]

Often we see that the initial reaction to an adverse event is *if only people would just abide by the rules, take care and be professional...*. It is fine to have an initial feeling of anger, and to have to count to ten before you are able to open up to why things made sense to people at the time. This is a human reaction and does not need to be a reason for a restorative approach to fail.

6.6 Conclusion

Occasionally, we are confronted with an adverse outcome, even if we are proactive about safety. In this chapter, I have presented how we can learn from incidents and reinstate communication after an adverse event. I have discussed the willingness to report occurrences, how to investigate an adverse event and how to maximise what we learn from it. I have also discussed how we might restore relations that have been damaged in the event.

Key actions:
- Have a satisfactory reporting system in place.
- Conduct an appropriate, independent investigation into an adverse event that avoids common pitfalls and applies language conducive to learning.
- Develop recommendations with stakeholders, and make them accountable for implementation.
- Check the mitigations for effectiveness.
- Regularly prune the risk log/action list to reflect real risks.
- Apply a restorative approach in an adverse event; consider appointing a coordinator.
- Inhibit the urge for disciplinary measures (including suspensions and investigating individual culpability) as a result of an adverse event.
- Recognise the need for all victims to determine the pace of the restorative approach themselves.
- Convey the demanding nature of Restorative Practice to sceptics.

The next chapter is about tailoring your approach to the safety level of your industry and your organisation.

Notes

1. ... the introduction of new technologies...: Mider, Z. (2019). Tesla's Autopilot Could Save the Lives of Millions, But It Will Kill Some People First. Bloomberg Businessweek, October 9, 2019. https://www.bloomberg.com/news/features/2019-10-09/tesla-s-autopilot-could-save-the-lives-of-millions-but-it-will-kill-some-people-first, accessed April 17th, 2020.

   ... societal developments...: Several 'Project X' riots have taken place due to invitations for parties going viral on social media. In one such case in the Netherlands, for € 250,000 damages were caused in the small northern town of Haren by vandals from all over the country. The ensuing investigation into the causes of the incident cost much more than that at € 400,000 but this was probably worth it because of the new insights that were drawn (and that prompted the mayor resign). See https://nl.wikipedia.org/wiki/Project_X_Haren

2. Dutch Safety Board (2011, p. 50).

3. ... we rely on those who are directly involved...: Our cognition doesn't always protect us from errors. In a laboratory study using an Airbus A320 fixed base flight simulator, de Boer, Heems and Hurts (2014) found long delays for pilots to identify a malfunction on the auto throttle even though multiple indications for the malfunction were in full view. In these experiments, the participant was the Pilot Flying, and he or she was accompanied by a researcher who did not support the identification of the failure. The manipulation consisted of an auto throttle malfunction of the left engine that was fixed at idle power. Contrary to what the pilots might expect, no warning messages appeared on the lower part of the upper Electronic Centralized Aircraft Monitor (ECAM). However, five indications were available to signal the malfunction: (1) a discrepancy between the primary axis speed indications of the left and right engines; (2) deviations of the exhaust gas temperature; (3) rotation speed of the secondary axis; (4) deviation in fuel flow between the two engines; and (5) rudder deflection indication, which was presented on the middle console next to the rudder trim knob. These cues were in direct sight of the pilots while the malfunction was present and are part of the standard scanning cycle. Four participants (11%) detected the failure within 45 seconds, but 12 out of the 35 participants (34%) did not detect the failure within the total test time of 12 minutes. The time until failure detection was generally shorter for more experienced pilots. In the simulator study, the distribution for the more experienced pilots showed the peak later in time, but the tail was less pronounced. That means that extreme short or long delays are less expected with experienced participants compared to inexperienced pilots, and that there is more intersubject consistency in the former group. Counterintuitively, the probability of tactic change did not follow a symmetrical distribution. Rather than having an equal chance for a 'rapid' or 'slow' response around an average, there is a larger proportion of people that take a longer than 'normal' time to notice an error. 'Normal' in these cases is defined as the time that most people take, that is the mode of the distribution. In other words, there is a reasonable chance of a rapid response, but also a significant chance of a very delayed reaction to follow a unimodal log-log probability distribution. de Boer (2012) studied the time it took for participants to change their problem-solving tactics when the original instructions turned out to be incompatible with the required result. The distribution again matched a unimodal log-log probability distribution. Another famous example is the Gorilla in our Midst experiment by Simons and Chabris (1999). Note, additionally, that although two pairs of eyes see more than one pair (de Boer 2014), there is nevertheless an overreliance on double checking despite its omnipresence (Hewitt, Chreim & Forster 2016).

... there are error traps everywhere...: My personal favourite error trap that I found on the internet somewhere is:

What I if told you
You read the top line wrong?

Another dire example: The US Navy is replacing touch screen controls on destroyers, after the displays were implicated in collisions. Unfamiliarity with the touch screens contributed to two accidents that caused the deaths of 17 sailors, said incident reports. Poor training meant sailors did not know how to use the complex systems in emergencies, they said. Sailors 'overwhelmingly' preferred to control ships with wheels and throttles, surveys of crew found. The US Navy reports looked into collisions involving the USS Fitzgerald in June 2017 and the USS McCain in August 2017. The Fitzgerald collided with a container ship near the Japanese mainland in an accident that killed seven sailors. The McCain was off the coast of Singapore when it hit a container ship, killing 10 of the Navy destroyer's crew. The incidents led to senior officers being charged with "negligent homicide". Others were dismissed from the service. Investigations found that both incidents were preventable and the result of "multiple failures". Strongly implicated in the collisions were the touch screen controls introduced on the destroyers. Service news website USNI reported that Rear Adm Bill Galinis, who oversees US Navy ship design, said the control systems were "overly complex" because shipbuilders had little official guidance on how they should work. As a result, he said, the control systems on different ships had little in common, so sailors often were not sure where key indicators, such as a ship's heading, could be found on screens. In addition, he said, a fleet survey about attitudes to the display-driven controls was 'really eye-opening'. 'We got away from the physical throttles, and that was probably the number one feedback from the fleet - they said, just give us the throttles that we can use,' said Rear Adm Galinis. The survey showed a desire for wheels and throttles that, prior to the introduction of touch screens, were common across many different types of vessel. The US Navy was now developing physical throttle and wheel systems that can replace the touch screens, USNI said. The service plans to start the process of replacing touch screens in the summer of 2020.

   BBC. (2019). US Navy to ditch touch screen ship controls, August 12th, 2019. https://www.bbc.com/news/technology-49319450, accessed August 17th, 2019.
4. … the willingness of front-line employees…: Remember the case of the building that was mistaken for the runway at Dublin (Chapter 3)? The crew did not even report the missed approach, thinking it was an insignificant error on their part. In this case, the modifications to the lighting of the building would not have been made if it wasn't for the attention of the media – despite a reporting system that was in place.
5. … The reporting system is…: Piric, Roelen, Karanikas, Kaspers, van Aalst and de Boer (2018).
6. Passenier, de Boer, Karanikas, Roelen, Ball, Piric and Dekker (2018).
7. … Overall, the reporting policy…: Karanikas, Soltani, de Boer and Roelen (2016); ICAO (2013a).
   … Professionals tend to prefer…: Hewitt and Chreim (2015).
8. Five good (and free at time of writing) resources to learn from events are: European Safety Reliability and Data Association (ESReDA) (2009, 2015); Allspaw, Evans and Schauenberg (2016); McLeod, Berman, Dickinson, Forsyth and Worthy (2020).
9. … is to understand…: Many organisations are now rebranding an accident investigation (that presupposes root causes) into *learning reviews*. Accident investigation techniques have remained essentially the same for many decades, yet the recognition that complexity is increasing in most organisations demands an added form of inquiry. The Learning Review, first adopted by the U.S. Forest Service, explores the human contribution to accidents, safety,

and normal work. It is specifically designed to facilitate the understanding of the factors and conditions that influence human actions and decisions by encouraging individual and group sensemaking at all levels of the organization. See: Pupulidy and Vesel (2017).

10. ... Pitfalls include...: Karanikas, Soltani, de Boer, Roelen, Dekker and Stoop (2015).

    · See also ESReDA (2009, 2015); Allspaw, Evans and Schauenberg (2016).

11. Air Accident Investigation Branch (2019).

12. ... methods that support the analysis...: These methods include Systems-Theoretic Accident Model and Processes (STAMP, Leveson 2011; Leveson & Thomas 2018), Functional Resonance Analysis Method (FRAM, Hollnagel 2017) and Accimap (Rasmussen 1997). These models are being promoted by followers, and they tend to ignore that the analysis methods that are incorporated in the models can also be applied independent of the modelling prescriptions. A key (supposed – not scientifically validated) benefit of the use of these models in an organisation is the resultant adoption of a system-wide perspective on safety.

13. ... The word 'finding' is often used...: A finding is the discovery and establishment of the facts of an issue. The word 'discovery' suggests that the investigator needs to actively pursue relevant findings. 'Establishment' suggests that credibility for the finding needs to be constructed before it can be interpreted as a fact. The phrase 'of an issue' suggests that there will be findings that are not relevant for understanding the event under consideration and should therefore be discarded. Findings can be facts that have been verified, presumptions that have been proven to be true or false (based on the available facts or analysis), or judgements beyond reasonable doubt when dealing with human and organisational factors. See: ESReDA (2009, p. 33).

14. ... to recognise the rationale...: If needed, revisit the section and notes on making sense of the situation in Chapter 2, or error traps in the introduction to this chapter.

    ... difficult to get into someone's head...: Morieux and Tollman (2014, p. 185), suggest that we don't need to apply pseudo-psychology and invoke people's mentality or mind-set. Instead, focus on how goals, resources and constraints explain the behaviour.

15. ... the ultimate question of life, the universe, and everything...: Adams (1979) (of course): The Hitchhiker's Guide to the Galaxy.

16. ... The language...: Dekker (2014b); Karanikas, Soltani, de Boer and Roelen (2015).

    ... folk models...: Dekker and Hollnagel (2004). Other often used folk models include 'complacency' and 'situational awareness'. See, for instance, the Irish Health and Safety Review, January/February 2020, p. 2: 'HAS warns against complacency as fatalities increase'.

17. The examples are taken from the investigations of the Dutch Safety Board into the crash of Turkish Airlines at Schiphol Airport in February 2009 and an aircraft taking off from a taxiway at Schiphol in February 2010.

    Dutch Safety Board (2010, p. 82); Dutch Safety Board (2011, pp. 6, 7, 43, 50, 93).

18. ... there is no 'correct' or absolute set of recommendations...: Contrary to what we might hope, there is no guarantee that the actions that we take (if any) are going to prevent similar incidents from happening again. In fact, Martin-Delgado and colleagues (2020) suggest 'Despite the widespread implementation of [incident investigations in health care] in the past decades, only two studies

[out of 21] could, to some extent establish an improvement of patient safety due to [these investigations]'. See also Karanikas, Soltani, de Boer and Roelen (2015); Drupsteen and Guldenmund (2014).

19. … the really big safety payoff…: Seymour, J. (2008). Canadian Aviation Safety Seminar 2008 Opening Remarks.
20. … the most valuable and difficult part of an investigation…: See, for instance, Card, Ward and Clarkson (2012); Kwok, Mah and Pang (2020); Parker, Ummels, Wellman, Whitley, Groeneweg and Drupsteen-Sint (2018).

    … Recommended actions can vary in effectiveness…: See Hollnagel (2008).

    … administrative improvements […] are four times more common…: Card and colleagues (Card, Ward & Clarkson 2012) report that of all the risk controls implemented as a result of root-cause analysis in hospitals that 78% were administrative. Improvements that were other than administrative were 1.6 times more likely to have explicitly reported success. Kwok and affiliates (Kwok, Mah & Pang 2020) show that 82% of the recommendations support 'weak' actions such as double checks, warnings and labels, new procedure/memorandum/policy, training and education (including counselling) or additional study/analysis. Interestingly, the authors themselves (p. 8), in an effort to improve the type of recommendations following a root-cause analysis, suggest only 'weak' actions: 'Conducting regular root cause analysis training, implementing easy-to-use root cause analysis tools, inviting members with human factors expertise to the root cause analysis panel, promoting a safety culture to staff in all public hospitals and aggregating analysis of incidents, to be possible actions adopted by the organisation's management'.

    … some 'improvements' actually complicate operations…: Allspaw, Evans and Schauenberg (2016).
21. … We often encounter…: For example, oil & gas industry, own experience, 2018; English airport, own experience 2020.
22. … create economic benefits…: Kaur, de Boer, Oates, Rafferty and Dekker (2019). See for more details of Restorative Practice: Dekker (2016, 2018); Dekker and Breakey (2016): 'Just Culture', the movie on http://www.safetydifferently.com.

    Kleinsman and Kaptein (2017) have collected case studies in Dutch mental healthcare where a restorative approach has been applied.
23. … traditional 'Just Culture'…: see notes in Chapter 2; Braithwaite (2004).
24. Weeda (2018a); Weeda, (2018b); UMC Utrecht (2019); Weeda (2020).
25. … A restorative approach is invoked…: The process that I describe is based on our own interpretation of the restorative checklists by:

    - Sidney Dekker (undated): Restorative Just Culture Checklist. Available at https://www.safetydifferently.com/wp-content/uploads/2018/12/RestorativeJustCultureChecklist-1.pdf, accessed April 20th, 2020.

    - Mersey Care NHS Foundation Trust (2020): Supporting Just and Learning Culture. March 2020, version 3. Proprietary, used in conjunction with the short course by Mersey Care NHS Foundation Trust and Northumbria University on Transforming Organisational Culture - Principles and Practise of Restorative 'Just Culture', see https://www.northumbria.ac.uk/study-at-northumbria/continuing-professional-development-short-courses-specialist-training/restorative-just-culture/

26. It has been suggested that remorse may be feigned, but if you can fake remorse and get away with it, then there is no real issue. Adapted from Hudson (2007, p. 714). Cognitive dissonance ensures that people who are that successful at faking start to believe their own speeches.

27. ... this approach is too lenient...: Restorative Practice has sometimes incorrectly been likened to no-blame culture: 'A no-blame culture recognises that operating problems are often complex and usually are not the fault of only one person. So where an operating problem has occurred, the business is far better off finding out what happened. The first step to achieving this is to stop looking for the culprit. Though there may be the odd person who comes to work intending to fail, the vast majority of us intend to succeed and as managers, our job is to help them, not discipline them' Hand (2016, p. 2). However, such a no-blame culture disregards the obligation that practitioners have to restore relations and distil learning from an incident and is therefore an inferior alternative to Restorative Practice. Berlinger (2005, p. 58).

28. ... In that case...: Personal discussion with Sidney Dekker and Nico Kaptein, August 19th, 2019.

7

Taking Action

7.1 Introduction

In this book, I have presented a new paradigm of safety that focuses on how to integrate the natural variability of human performance – and our ability to compensate for variability elsewhere – into organisations to enable successful outcomes. An increasing number of companies are embracing these thoughts. Your organisation can likewise benefit if you take the lead and initiate action. Each of the preceding chapters was summarised with key points that will help you implement safety leadership. I suggest that you initially adopt them in the order that they were written, although soon you will find that the key points are mutually supportive – within and across chapters.[1]

Our approach has been applied in many different domains. We have seen that the emphasis of this method needs to be adapted to suit local circumstances, building on the safety processes that are already embedded in the organisation and the (moral) values that are being adhered to. We also need to take the risk level of the industry as a whole into account. Below I discuss five different clusters of industries: ultra-safe, process, construction, healthcare and military. I also discuss the application of our approach to regulatory bodies.[2]

7.2 Ultra-safe Industries

Commercial aviation and other ultra-safe industries like nuclear power and European railroads have achieved a safety level of around one accident per 100,000 or even a million events (e.g. departures or trips). Near-misses and accidents occur as a result of interactions, not due to a broken component or a single action. The sparse incidents that do occur are less predictive of future events than in other industries. Despite this, the natural inclination after adverse events is to add rules and procedures in an effort to avert a similar occurrence, leading to overregulation. Safety burdens are added as a matter of course without a distinguishable safety effect. Thinking in terms of linear safety barriers needs to be replaced by models that are appropriate for complex environments. The recommended focus in these industries is by recalibrating the balance between *paper* and *procedures*, and reducing safety procedures and bureaucratic compliance (perhaps using micro-experiments). The discussion on risk needs to be kept alive despite the absence of incidents and operators need to gain experience at the edges of the operating envelope.[3]

7.3 Process Industry and Road Infrastructure

The oil and gas industry and chemical industry represent a risk of between one accident in a thousand events to one accident in 100,000 events. Examples of other industries with a similar safety record include road traffic and some better-performing domains within healthcare such as anaesthesiology and blood transfusions. In these domains, drift manifests itself as the infrastructure ages and the assumptions behind design choices are lost or become outdated. Numerous examples have already been mentioned such as the fires in the ExxonMobil refinery in Rotterdam (paragraph 5.2.1) and at the chemical process plant in Moerdijk, the Netherlands (paragraph 5.2.3), and the gap between the prescribed isolation of process equipment and current practice (paragraph 5.2.5). A bridge over the Merwede River in the Netherlands was shut immediately for trucks in October 2016 after an investigation found that the bridge was much weaker than expected and about to collapse – much like the Morandi Bridge in Genoa collapsed suddenly in August 2018. A focus on bureaucratic compliance means that design assumptions are not monitored and that unmanaged gaps between *paper* and *practice* abound. The emphasis should be on maintaining an alignment between assumptions and reality. Previous incidents and near-misses can point to the riskiest gaps, so an effective reporting system is important. There are often opportunities for error-proofing of designs through micro-experiments.[4]

7.4 Construction Industry

The construction industry – like the process industry – represents a risk of between one accident in a thousand events to one accident in 100,000 events. Here, the creation (if not yet available) and improvement of rules and procedures can help to advance safety. Again a focus on bureaucratic compliance means that unmanaged gaps between *paper* and *practice* abound. Much can be learned from previous incidents and near-misses, so an effective reporting system is important. Designs can be error-proofed and standardised through micro-experiments.[5]

> On the morning of January 11th, 2017, the Polish construction worker Pawel Kupczałojć is found dead at a construction site in Utrecht, the Netherlands. He seems to have fallen down an unprotected elevator shaft from 20 meters up. There was no barrier, no warning sign and even no lighting on the fourth floor at the time that Pawel fell – a makeshift barrier was installed later that morning after he was found.[6]

Error-proofing and reports of this dangerous situation are hopefully effective in avoiding future mishaps.

7.5 Healthcare

Most parts of the healthcare system have a safety record of one accident in a hundred to a thousand events. This is similar to the error rate for individuals - and thus represents a safety level that is typical of craftsman-like professions. To enable improvement and stay affordable, healthcare needs to identify and implement patient-focused processes that are aimed at improving overall wellbeing. A process-orientated healthcare system is better able to identify factors that frustrate or promote excellence, to disseminate learnings, to maintain competences, to introduce critical checkpoints and to regulate supply and demand. Standardised processes that cut across disciplines and hierarchies and focus on creating value for care-users need to be defined. The creation, implementation and continuous improvement of procedures would help to sustain these standardised processes. In many parts of healthcare, a process approach is possible, representing a step change in progress towards safety and efficiency. However, often the attempts that have been made to transition healthcare to a more process-orientated industry have had a limited effect, in part due to the incumbents' interests to protect their autonomy.[7]

> Recently my father was in hospital and recovering from an operation. A nurse came up to us (as promised) to tell us the successful outcome. She added (priding herself) that she had actually gone off duty but thought it important to let us know how my father was doing. As much as we appreciated her effort, little did she realise that she robbed the system of a learning opportunity by compensating for the flaws in the communication process of the hospital. Rather than relying on off-duty nurses to inform the next-of-kin, the hospital needs to consider how to deliver a post-surgery debrief consistently and effectively.[8]

Medical equipment – but particularly processes – can be error-proofed and standardised through micro-experiments. Much can be learnt from previous incidents and near-misses, so an effective reporting system is important.

Improvements such as checklists or a training on teamwork that are not embedded in improved processes generate perfunctory results. Currently more than in other industries, individuals are held personally culpable for adverse events, which understates other contributing factors, inhibits openness and truthfulness, and reduces the opportunity for learning. It is therefore important that following an adverse event, that every effort is made to reconcile relations and support practitioners using a restorative approach.[9]

7.6 Military

The armed forces seem different to the other industries in that danger is sought rather than avoided. Services across the globe are marked by an eagerness to accomplish a mission – whatever it takes – which is often characterised as a *can-do mentality*. This attitude values goal realisation at the expense of a large gap between *paper* and *practice*, and is amplified by austerity and uncompromising demands from politics (or a wish by generals to be relevant to society).

> In March 2016 the Royal Dutch Army's Commando Corps was exercising on a firing range at the Police Academy in Ossendrecht. During the exercise the trainees used live ammunition on a firing range that did not have bulletproof walls. An instructor that was trailing the attack force was invisible behind one of the walls. He was shot through this wall and fatally injured as the attack force cleared the room of enemies. Besides the lack of an appropriate facility, shortcomings were evident in the procedures and regulations for the exercise location, the qualifications of the instructors and in the design of the counterterrorism training programme. The Dutch Safety Board noted in its report that the army prioritised the timely training of personnel and the scheduled progress of exercises, without taking heed of the consequences of an accumulation of shortcomings in safety.[10]

Other examples of non-hostile losses include the fire aboard a RAF Nimrod maritime patrol aircraft in Afghanistan in 2006, two collisions involving the Japan-based US Seventh Fleet in 2017, and the unintentional and premature detonation of a mortar round of the Dutch Airborne Brigade in Mali in 2016. It is acknowledged that risks are to be expected during operational missions and may also be present during exercises. However, the callous pursuit of operational deployment despite a tangible lack of resources, and the creation of significant gaps between *practice* and *paper* whenever an objective is jeopardised, are clear signs that safety margins are eroding and that barriers have disintegrated. Military leaders are encouraged to conduct a review of gaps between *paper* and *practice* whenever a mission is fulfilled rather than to celebrate this as another accomplishment. This makes it possible to identify whether unjustified risks were taken (and we got lucky) or whether the exception was tolerable. Rules and regulations need to take these routine concessions into consideration (creating *freedom in a frame*), where possible through micro-experiments. These dispensations (*freedom*) – and their corresponding hard constraints (*the frame*) – require approval and enforcement not just at the operational level but also at the tactical and strategic level. Much will need to be invested at all levels in maintaining the alignment between *paper* and *practice* and getting exceptions reported. Procedures – but also hardware – can be error-proofed by involving operators in the design process and utilising user feedback.[11]

7.7 Regulatory Bodies

Regulators can play an important role in encouraging organisations to advance safety by adopting the principles described in this book. As a regulator, rather than hunt for compliance, I would want to know whether a company really understands how it gets work done. I would find out how managers ensure that they hear bad news. They need to have a good understanding of where work is difficult and what sacrifices are being made in the pursuit of multiple goals. Processes cut through organisational and hierarchical boundaries and are aimed at safely creating value. I would like to see evidence of a collaborative effort to close gaps. I would like to hear how management and staff maintain alignment between Work-as-Done and Work-as-Imagined, and how exemptions are followed up. I would like to see evidence of deliberations to find the right balance between *freedom* and the *frame*. I would look to see how new innovations are introduced in a controlled and safe-to-fail manner and monitored for unexpected outcomes. It would be great to see proof of diversity, discussions on risks and operators that are challenged to gain experience at the edges of the operating envelope. Of course, I would expect to see that learning arises from incidents, and evidence of efforts to restore relations rather than to seek retribution. I would refrain from commenting about a (lack of) safety culture, and certainly not require the company to expand or harshen its safety rituals that only serve to distract from the real issues. And as far as one's own enforcement powers go, I would seek to create a high probability of detection of the shortcomings mentioned above, instead of only responding to incidents. I would also ensure that among different regulatory bodies, we cover the whole playing field. We don't want new technologies to be available without oversight because 'it isn't in our scope'. Any penalties that we administer should be independent of the consequences of the event – and we need to counter pressure from public, politics and press to the contrary.[12]

7.8 Conclusion

In this book, I have described a coherent approach to safety that is different, doable and directed. This approach is suitable for organisations that are subject to a low incident rate where adverse events are too few to count and their cause is not a single, easily identifiable failure or malfunction. Incidents are a natural consequence of the complexity of the system, whose effect we can diminish if we know what's going on and are able to react appropriately.

I have described an approach that is *different* to traditional safety thinking. We ask *what* caused an event, not *who*. So-called 'human errors' are a symptom of underlying causes, such as system design, training, supervision or conflicting goals. I have emphasised that procedures are not always up to date or appropriate, and we need to understand how work really gets done, even when conditions are poor and resources are slim, rather than stalking after compliance. An absence of adverse events doesn't necessarily mean that safety margins are preserved, and the pursuit of numerical goals leads to manipulations that frustrate transparency and understanding. I have advocated an alternative to a traditional 'Just Culture' in the wake of adverse events to cure hurt and restore relations.

I have offered an approach that is *doable*, by showing the steps to improve and maintain safety. I have suggested how to identify where safety is challenged and how to address those issues. I have demonstrated a sensible approach to interventions, and how to monitor their effect on safety and performance. I explain how to recognise eroding safety margins and how to counter them. And I have included what to do when things go wrong, both to maximise learning and to restore the feelings of hurt and pain.

This approach needs to be *directed*. Teams need the freedom to be curious without passing judgement, finding out about how work is really done without referring back to how the task is described on paper or how the task 'should' have been done. You need to empower people to tell you about gaps between the procedures and how they actually do their work, and sometimes you need to keep press, public, politics and even their peers at bay.

I wish you a safe journey. You are ready to take the first steps.

Notes

1. ... an increasing number...: For instance in 2019, Netflix advertised for a Senior
   Resilience Engineering Advocate, 'to improve reliability and resilience of Netflix
   services by focusing on the people within the company, since it's the normal

everyday work of Netflix employees that creates [Netflix's] availability. To cultivate operational excellence, [the Resilience team] reveal risks, identify opportunities, and facilitate the transfer of skills and expertise of [Netflix] staff by sharing experiences.' https://jobs/netflix.com/jobs/869465, accessed March 2nd, 2020.

2. ... and the (moral) standards...: for instance, the 'can do' mentality of the military, the compliance mind-set in oil & gas, and the cherished autonomy of medical specialists.

3. ... by models that are appropriate...: Models that are appropriate for ultrasafe and therefore complex environments include Systems-Theoretic Accident Model and Processes (STAMP, Leveson 2011; Leveson & Thomas 2018), Functional Resonance Analysis Method (FRAM, Hollnagel 2017) and Accimap (Rasmussen 1997). See also ICAO (2019); Amalberti (2001).

4. ... In these industries...: Amalberti (2001); Amalberti, Auroy, Berwick and Barach (2005).

   ... A bridge over the Merwede River...: Demoed, K. (2019). Hoogleraren: 'Merwedebrug is op haar na ingestort'. Eenvandaag, January 17th, 2019. https://eenvandaag.avrotros.nl/item/hoogleraren-merwedebrug-op-haar-nais-ingestort/, accessed June 29th, 2020.

5. ... construction...: In the Netherlands, 20 people died in the construction industry in 2018. About 450,000 people are employed in this sector, making the fatalities 1 in 22,500. (https://www.inspectieszw.nl/actueel/nieuws/2019/02/26/ inspectie-szw-roept-bedrijven-op-meer-te-investeren-in-veiligheidscultuur; https://nl.wikipedia.org/wiki/Bouw).

6. NRC. (2019). Voor het gat was dichtgetimmerd viel Pawel dood. NRC Handelsblad, Amsterdam, July 10th, 2019. https://www.nrc.nl/ nieuws/2019/07/10/voor-het-gat-was-dichtgetimmerd-viel-pawel-dooda3966670, accessed August 5th, 2019.

7. ... typical of craftsman-like professions...: A recent study suggested that the error rate across healthcare in the United States is one fatality by medical error per 150 hospitalisations (Makary and Daniel 2016), but this seems inflated. Sunshine, Meo, Kassebaum et al. (2019) suggest a rate about half of this for all adverse events of medical treatment (AEMT), so not just medical errors but also adverse drug events, surgical and perioperative adverse events, adverse events associated with medical management, and adverse events associated with medical or surgical devices. Nevertheless, these fatality rates are also typical of bungee jumping, avalanche releasing and mountain climbing. Not surprisingly, the human error rate for individuals is approximately 10^{-3} so in a sequential craftsman-like healthcare process, the error/death rate is a multiplication of that number by the number of steps in the process, mitigated by both the probability of timely discovery and the chance that the error is not fatal. The craftsman-like organisation of healthcare is visible through:

- Personal contact between care-user and the medic.

- Taking an oath.

- Bespoke processes/treatment.

- Personal pride in and culpability for end result.

- Master – apprenticeship training on the job and strong hierarchy dependent on seniority.

- A self-employed status (valid for GPs generally and for specialists in some regions, for example the Netherlands for non-academic hospitals).
- Strong guild system.
- Support staff may be numerous but with limited authority.

… To improve safety…: Patterson and Wears (2015) describe how the pressures on a hospital pharmacist lead to adaptations in the way that work is done to accommodate these. These adaptations lead to successful outcomes although the authors describe how safety margins have all but eroded. Their recommended proactive solution is to ensure 'changes [...] occur relative to how and with whom information is shared as well as a basic examination of the work and how the work is structured and accomplished in the system' (i.e. create value-focused processes across disciplines) and to ensure 'additional resources are deployed when the work exceeds a certain level'. They describe how the pharmacist's supervisor has some understanding of the challenges but is constrained by the current system.

… attempts have been made…: Attempts have been made to transition healthcare from a craftsman-like profession to a more process-orientated industry, but so far these have had a limited effect. Healthcare suffers from what has been called 'excessive actor autonomy', implying that the independence of the healthcare professionals needs to be curtailed for progress to be made to the next safety level where 'equivalent actors' are interchangeable and able to provide a similar service rather than counting on each specific individual to deliver their own variant. The transition to a standardised system is impeded by the incumbents' interests, who are effective in mobilising care-users to promote their aims. Who doesn't want the most senior surgeon to execute the operation, to do the intake himself or herself, and to tell us the results? We even have patient websites that offer reviews of individual healthcare professionals. The process approach is probably not perceived to be as consumer-friendly as an individual and bespoke treatment, but it is a lot cheaper and safer than the approach of a craftsman (and which both help satisfy consumer interests in different and significant ways).

In Holland, many dentists have taking the service/non-identity approach already, by utilising multiple chairs per professional in which care-users are prepared for the examination and their history is noted. The dentist dashes in for the inspection and diagnosis, and off he or she is again to the next chair. Larger treatments are scheduled separately, requiring you to come back for a second appointment. Clinics exclusively for routine therapy like laser treatment of eyes, certain types of skin treatment or orthopaedic surgery have sprung up and are easily more competitive than the regular institutions. To be fair, many parts of the healthcare system have been proceduralised, but the overarching theme is still one of individual heroes. I don't think that Richard de Crespigny (who incidentally is an ambassador for safety at Australia's largest hospital) or Sully Sullenberger consider themselves heroes despite being instrumental in safely bringing down very crippled aircraft. They know that they are indebted to fellow-crew members, aerospace designers and engineers and air traffic control to name a few. Amalberti (2001); Amalberti, Auroy, Berwick and Barach, (2005);

De Vos (2018); Semrau, K. E., Hirschhorn, L. R., Marx Delaney, M., Singh, V. P., Saurastri, R., Sharma, N., ... Kodkany, B. S. (2017). https://www.zorgkaartnederland.nl/, accessed April 4th, 2020.

Wexler, A. (2020). The trouble with calling health care workers 'heroes'. *The Boston Globe,* April 10th, 2020. https://www.bostonglobe.com/2020/04/10/opinion/trouble-with-calling-health-care-workers-heroes/, accessed April 17th, 2020.

Amust read on the lack of processes in healthcare (although it is difficult to verify the truthfulness of the narratives) is Kay (2017).

8. Own experience at Isala hospital, Zwolle, the Netherlands, in June 2019.

9. ... individuals are held personally culpable...: Young doctors in the Netherlands felt the need to organise an 'Error Festival' to help each other cope with their 'individual mistakes'. Interestingly, in healthcare, physicians can sometimes not see their patient's suffering, but rather only the impact of the error on his career, income and prestige. Rituals to forge forgiveness after an error are between medics, with the injured party (the first victim) excluded. The erring medic's superior forgives on their behalf. Berlinger (2005, pp. 49, 89).

Dagblad, A. (2018). Jonge artsen bespreken medische missers, November 26th, 2018. https://www.ad.nl/binnenland/jonge-artsen-bespreken-medische-missers~adaf21bd/, accessed April 17th, 2020.

10. Dutch Safety Board. (2017). Lessons from the Ossendrecht shooting incident. Available at https://www.onderzoeksraad.nl/en/page/4293/lessen-uit-schietongeval-ossendrecht, accessed March 3rd, 2020.

11. ... Other examples...: Mc Garth. (2017). What we learned from the Navy's collision inquiries. Blog on War on the Rocks, https://warontherocks.com/2017/11/what-we-learned-from-the-navys-collision-inquiries/, accessed March 3rd, 2020.

Army Technology. (2019). Conference sets safety at top of defence agenda. Available at https://www.army-technology.com/features/safety-in-defence-conference/, accessed March 3rd, 2020.

Dutch Safety Board. (2017). Mortar Accident Mali. Available at https://www.onderzoeksraad.nl/en/page/4401/mortierongeval-mali, accessed March 3rd, 2020.

... the callous pursuit...: 'A general explains: our can-do mentality is our trap. Every time we say: 'we make it happen'. And we reduce resources even further, and you swear, but a day later you are back in the mode of it needs to happen, so it will. But the noose becomes tighter and tighter around your neck. So, we are stuck in our can-do mentality and our loyalty.' Tesselaar and Rodermond (2017, p. 99).

... lack of resources...: We studied the gaps between rules and actual work at a helicopter maintenance facility of the Dutch military. We were able to highlight the large size of the gap between Work-as-Imagined and Work-as-Done, and a mentality in the organisation to 'get stuff done' despite a desperate lack of resources. We called this a 'Can Do' mentality and were able to mirror that across the Dutch and even international military.

Kurtz (2014); de Boer (2016); Kapetanovic. (2017). *Safety attitudes at a maintenance squadron using 'narratives'*, BSc thesis, Amsterdam University of Applied Sciences, 2017, [in Dutch].

12. ... cover the whole playing field...: In September 2018, four children under nine years old died near Oss, the Netherlands, when an electric multi-person bicycle (called 'Stint') was unable to stop for a railway crossing and was hit by a passing train. The vehicle turned out to have multiple technical issues but its market entry was not supervised by the regulatory bodies (RDW and ILT) due to unclarity about each of their scope. De Telegraaf (2020). Ongeluk met de Stint was onnodig, January 18th. Available at https://www.telegraaf.nl/financieel/3036680/ongeluk-met-de-stint-was-onnodig, accessed July 8th, 2020.

References

Air Accident Investigation Branch. (2019). AAIB Bulletin S1/2019. Aldershot, UK. Available at https://assets.publishing.service.gov.uk/media/ 5c73c02bed915d4a3d3b2407/S1-2019_N264DB_Final.pdf

Air Accident Investigation Branch. (2019). AAIB Bulletin S2/2019. Aldershot, UK. Available at https://assets.publishing.service.gov.uk/media/ 5d53ea15e5274a42d19b6c2e/AAIB_S2-2019_N264DB.pdf

Allspaw, J., Evans, M., & Schauenberg, D. (2016). *Debriefing facilitation guide.* Etsy, Brooklyn, NY. Available at https://extfiles.etsy.com/DebriefingFacilitationGuide. pdf, accessed June 21st, 2020.

Allwood, C. M. (1984). Error detection processes in statistical problem solving. *Cognitive Science*, 8(4), 413–437.

Amalberti, R. (2001). The paradoxes of almost totally safe transportation systems. *Safety Science*, 37(2–3), 109–126.

Amalberti, R., Auroy, Y., Berwick, D., & Barach, P. (2005). Five system barriers to achieving ultrasafe health care. *Annals of Internal Medicine*, 142(9), 756–764.

Amalberti, R., & Vincent, C. (2020). Managing risk in hazardous conditions: improvisation is not enough. *BMJ Quality & Safety*, 29, 60–63. doi:10.1136/ bmjqs–2019–009443.

Amernic, J., & Craig, R. (2017). CEO speeches and safety culture: British Petroleum before the Deepwater Horizon disaster. *Critical Perspectives on Accounting*, 47, 61–80.

Anand, N. (2015a). Breaking the myth: the effectiveness of bowties in risk and safety management. *BIMCO Bulletin*, 110(3), 34–35, Bagsværd., Denmark.

Anand, N. (2015b, June). Risk assessment at the sharp end. *Seaways, Journal of the Nautical Institute*, London.

Anand, N. (2020). Do we really need a just culture? Presentation at the Institution of Occupational Safety and Health (IOSH) Ireland South Branch event Start a ripple, create a wave, Cork, February 2020.

Australian Transport Safety Board. (2019). Incorrect configuration for landing involving Airbus A320, VH-VQK Ballina/Byron Gateway Airport, New South Wales, on May 18th, 2018. Canberra, Australia.

Bakx, G. C., & Richardson, R. A. (2013). Risk assessments at the Royal Netherlands Air Force: an explorative study. *Journal of Risk Research*, 16(5), 595–611.

Barnett, A., & Wang, A. (2000). Passenger mortality risk estimates provide perspectives about flight safety. *Flight Safety Digest*, 19(4), 1–12.

Beckers, J. (2019). Process improvements with safety differently. BSc thesis, Aviation Academy, Amsterdam University of Applied Sciences.

Berlinger, N. (2005). *After harm: medical error and the ethics of forgiveness (Vol. 18)*. Baltimore, MD: Johns Hopkins University Press

Berlinger, N. (2016). *Are workarounds ethical? Managing moral problems in health care systems*. Baltimore, MD: Oxford University Press.

Braithwaite, J. (2004). Restorative justice and de-professionalization. *The Good Society*, 13(1), 28–31.

Brown, S. (2019). The London Luton airport safety differently journey. Available at https://safetydifferently.com/the-london-luton-airport-safety-differently-journey/, accessed January 23rd, 2020.

Bureau d'Enquêtes et d'Analyses pour la sécurité de l'aviation civile. (2019). Investigation Report Serious incident to the AIRBUS A340-313E registered F-GLZU and operated by AIR FRANCE on 11 March 2017 at Bogotà (Colombia). BEA2017–0148, Paris.

Busch, C. (2019). Heinrich's Local Rationality: Shouldn't 'New View' Thinkers Ask Why Things Made Sense To Him? MSc thesis, Lund University, Sweden.

Caprari, E., & Van Wincoop, A. (2020). Veilig gedrag bij beweegbare bruggen: Het gebruik van nudges om veilig gedrag te stimuleren. *Tijdschrift voor Human Factors*, 45(2), 13–18 [in Dutch].

Card, A. J., Ward, J., & Clarkson, P. J. (2012). Successful risk assessment may not always lead to successful risk control: a systematic literature review of risk control after root cause analysis. *Journal of Healthcare Risk Management*, 31(3), 6–12.

Christensen, C. M., Hall, T., Dillon, K., & Duncan, D. S. (2016). Know your customers' jobs to be done. *Harvard Business Review*, 94(9), 54–62.

Clark, T. R. (2020). *The 4 stages of psychological safety: defining the path to inclusion and innovation*. Berrett-Koehler Publishers.

Cook, R. I., & Nemeth, C. P. (2010). "Those found responsible have been sacked": some observations on the usefulness of error. *Cognition, Technology & Work*, 12(2), 87–93.

Cromie, S., & Bott, F. (2016). Just culture's "line in the sand" is a shifting one; an empirical investigation of culpability determination. *Safety Science*, 86, 258–272.

de Boer, R. J. (2012). Seneca's error: an affective model of cognitive resistance. PhD thesis, Delft University of Technology.

de Boer, R. J. (2016). Research into control loop flaws at a maintenance squadron using narratives. In the *Proceedings of the 32nd EAAP Conference in Cascais*, Portugal, September 2016 (p. 153).

de Boer, R. J., Coumou, T., Hunink, A., & van Bennekom, T. (2014). The automatic identification of unstable approaches from flight data. In *6th International Conference on Research in Air Transportation, ICRAT* (pp. 26–30).

de Boer, R. J., & Dekker, S. W. A. (2017). Models of automation surprise: - results of a field survey in aviation. *Safety*, 3, 20.

de Boer, R. J., Heems, W., & Hurts, K. (2014). The duration of automation bias in a realistic setting. *The International Journal of Aviation Psychology*, 24(4), 287–299.

de Boer, R. J., & Hurts, K. (2017). Automation surprise: results of a field survey of Dutch pilots. *Aviation Psychology and Applied Human Factors*, 7(1), 28–41.

de Boer, R. J., Koncak, B., Habekotté, R., & Van Hilten, G. J. (2011). Introduction of ramp-LOSA at KLM Ground Services. In D. de Waard, N. Merat, A. H. Jamson, Y. Barnard, & O. M. J. Carsten (Eds.) (2012), *Human factors of systems and technology* (pp. 139–146). Maastricht, The Netherlands: Shaker Publishing.

De Crespigny, R. (2012). QF32. Macmillan, Australia.

De Vos, M. S. (2018). Healthcare improvement based on learning from adverse outcomes. Doctoral dissertation, Leiden University Medical Centre, The Netherlands.

Dekker, S. (2009). Report of the Flight Crew Human Factors Investigation for the Dutch Safety Board into the Accident of TK1951, Boeing 737-800 near Amsterdam Schiphol Airport, February 25, 2009. Lund University, School of Aviation, Sweden

Dekker, S. (2011). *Drift into failure: from hunting broken components to understanding complex systems*. Boca Raton, FL: CRC Press

Dekker, S. (2014a). *Safety differently: human factors for a new era*. Boca Raton, FL: CRC Press

Dekker, S. (2014b). *The field guide to 'human error'*, 3rd edition. London and New York: Routledge

Dekker, S. (2016). *Just culture: balancing safety and accountability*, 3rd edition. Boca Raton, FL: CRC Press

Dekker, S. (2017). *The Safety Anarchist: relying on human expertise and innovation, reducing bureaucracy and compliance*. London and New York: Routledge

Dekker, S. (2019). *Foundations of safety science: a century of understanding accidents and disasters*. Boca Raton, FL: CRC Press

Dekker, S. W., & Breakey, H. (2016). 'Just culture:' improving safety by achieving substantive, procedural and restorative justice. *Safety Science*, 85, 187–193.

Dekker, S., Cilliers, P., & Hofmeyr, J. H. (2011). The complexity of failure: implications of complexity theory for safety investigations. *Safety Science*, 49(6), 939–945.

Dekker, S., & Hollnagel, E. (2004). Human factors and folk models. *Cognition, Technology & Work*, 6(2), 79–86.

Deloitte. (2009). Mining Safety: A Business Imperative. Paper in the series Thought Leadership, Deloitte & Touche. Johannesburg, South Africa.

Dennett, D. C. (2013). *Intuition pumps and other tools for thinking*. London: Allan Lane.

Di Lieto, A. (2015). *Bridge resource management: from the Costa Concordia to navigation in the digital age*. Brisbane: Hydeas Pty Limited.

Drupsteen, L., & Guldenmund, F. W. (2014). What is learning? A review of the safety literature to define learning from incidents, accidents and disasters. *Journal of Contingencies and Crisis Management*, 22(2), 81–96.

Dutch Safety Board. (2010). *Crashed During Approach, Boeing 737-800, Near Amsterdam Schiphol Airport, 25 February 10th 2009*, The Hague.

Dutch Safety Board. (2011). *Take-off from Taxiway, Amsterdam Airport Schiphol*. The Hague.

Edmondson, A. C. (2018). *The fearless organization: creating psychological safety in the workplace for learning, innovation, and growth*. Hoboken, NJ: John Wiley & Sons

Elkind, P., Whitford, D., & Burke, D. (2011). BP: 'An accident waiting to happen'. *Fortune Features*, 85, 1–14.

European Safety Reliability and Data Association (ESReDA). (2009). *Guidelines for safety investigations of accidents*, ISBN 978-82-51-50309-9. Available at www.esreda.org.

European Safety Reliability and Data Association (ESReDA). (2015). *Guidelines for preparing a training toolkit in event investigation and dynamic learning*. Available at www.esreda.org.

Furness, K. (2020). All the way – the Maersk Safety Differently experience. Presentation at the Institution of Occupational Safety and Health (IOSH) Ireland South Branch event Start a ripple, create a wave, Cork, February 2020.

Gneezy, U., & Rustichini, A. (2000). A fine is a price. *The Journal of Legal Studies*, 29(1), 1–17.

Guldenmund, F. W. (2000). The nature of safety culture: a review of theory and research. *Safety Science*, 34(1–3), 215–257.

Hale, A., & Borys, D. (2013a). Working to rule, or working safely? Part 1: a state of the art review. *Safety Science*, 55, 207–221.

Hale, A., & Borys, D. (2013b). Working to rule or working safely? Part 2: the management of safety rules and procedures. *Safety Science*, 55, 222–231.

Hand, D. (2016). How to foster a no-blame culture. *The CEO Magazine*, September 9, 2016. https://www.theceomagazine.com/business/management-leadership/foster-no-blame-culture/, accessed April 12th, 2020.

Health and Safety Executive. (2019). *Understanding the impact of business to business health and safety 'rules'*. Report, London.

Henriqson, É., Schuler, B., van Winsen, R., & Dekker, S. W. (2014). The constitution and effects of safety culture as an object in the discourse of accident prevention: a Foucauldian approach. *Safety Science*, 70, 465–476.

Heraghty, D., Rae, A. J., & Dekker, S. W. (2020). Managing accidents using retributive justice mechanisms: when the just culture policy gets done to you. *Safety Science*, 126, 104677.

Hewitt, T., Chreim, S., & Forster, A. (2016). Double checking: a second look. *Journal of Evaluation in Clinical Practice*, 22(2), 267–274.

Hewitt, T. A., & Chreim, S. (2015). Fix and forget or fix and report: a qualitative study of tensions at the front line of incident reporting. *BMJ Quality & Safety*, 24(5), 303–310.

Hollnagel, E. (2008). Risk+ barriers= safety? *Safety Science*, 46(2), 221–229.

Hollnagel, E. (2009). *The ETTO principle: efficiency-thoroughness trade-off: why things that go right sometimes go wrong*. London: Ashgate Publishing, Ltd.

Hollnagel, E. (Ed.). (2013). *Resilience engineering in practice: A guidebook*. London: Ashgate Publishing, Ltd.

Hollnagel, E. (2017). *FRAM: The functional resonance analysis method: modelling complex socio-technical systems*. Boca Raton, FL: CRC Press.

Hollnagel, E. (2019). Synesis. Available at https://safetysynthesis.com/safetysynthesis-facets/synesis/, accessed June 27th, 2020.

Hollnagel, E., Nemeth, C. P., & Dekker, S. (Eds.). (2009). *Resilience engineering perspectives: preparation and restoration (Vol. 2)*. London: Ashgate Publishing, Ltd..

Hollnagel, E., Wears, R. L., & Braithwaite, J. (2015). *From Safety-I to Safety-II: A White Paper*. The resilient health care net: published simultaneously by the University of Southern Denmark, University of Florida, USA, and Macquarie University, Australia.

Hopkins, A. (2012). *Disastrous decisions. The human and operational causes of the Gulf of Mexico blowout*. Sydney: CCH Australia Limited.

Hudson. (2007). Implementing a safety culture in a major multi-national. *Safety Science*, 45(6), 697–722.

ICAO. (2013a). *Safety management manual*, 3rd edition. International Civil Aviation Organisation, Document 9859, Montreal.

ICAO. (2013b). Annex 19 to the Convention on International Civil Aviation - Safety Management. International Civil Aviation Organisation, AN19, ISBN 978-92-9249-232-8, Montreal.

ICAO. (2019). *ICAO Safety Report 2018 Edition*. International Civil Aviation Organization, Montreal.

Ioannou, C., Harris, D., & Dahlstrom, N. (2017). Safety management practices hindering the development of safety performance indicators in aviation service providers. *Aviation Psychology and Applied Human Factors*, 7(2), 95.

Johnson-Laird, P. N. (1983). *Mental models: towards a cognitive science of language, infer-ence, and consciousness (No. 6).* Cambridge MA: Harvard University Press.

Johnson-Laird, P. N. (2006). *How we reason.* Oxford, UK: Oxford University Press, .

Kahneman, D. (2011). *Thinking, fast and slow.* New York: Farrar, Straus and Giroux

Karanikas, N., Soltani, P., de Boer, R. J., & Roelen, A. (2015). Evaluating advancements in accident investigations using a novel framework. In the *Proceedings of the Air Transport and Operations Symposium 2015,* Delft (pp. 1–10).

Karanikas, N., Soltani, P., de Boer, R. J., Roelen, A., Dekker, S. and Stoop, J. (2015). *[Company] Project Report,* Aviation Academy, Amsterdam. Company Confidential.

Karanikas, N., Soltani, P., de Boer, R. J., & Roelen, A. L. (2016). Safety culture develop-ment: the gap between industry guidelines and literature, and the differences amongst industry sectors. In Arezes, P. (ed.), *Advances in safety management and human factors, Proceedings of the AHFE 2016 International Conference on Safety Management and Human Factors,* July 27–31, 2016, Walt Disney World®, Florida, USA, Springer (pp. 53–63).

Kaspers, S., Karanikas, N., Piric, S., van Aalst, R., de Boer, R. J., & Roelen, A. (2017). Measuring safety in aviation: empirical results about the relation between safety outcomes and safety management system processes, operational activi-ties and demographic data. In *PESARO 2017: The Seventh International Conference on Performance, Safety and Robustness in Complex Systems and Applications, IARIA,* (pp. 9–16). Red Hook, NY.

Kaspers, S., Karanikas, N., Roelen, A., Piric, S., & de Boer, R. J. (2019). How does avia-tion industry measure safety performance? Current practice and limitations. *International Journal of Aviation Management,* 4(3), 224–245.

Kaur, M., de Boer, R. J., Oates, A., Rafferty, J., & Dekker, S. (2019). Restorative just culture: a study of the practical and economic effects of implementing restor-ative justice in an NHS trust. In *MATEC Web of Conferences* 273(01007), 1–9. EDP Sciences, Les Ulis, France. doi: 10.1051/matecconf/201927301007

Kay, A. (2017). *This is going to hurt – secret diaries of a junior doctor.* London: Picador.

Klein, G. A., Orasanu, J. E., Calderwood, R. E., & Zsambok, C. E. (1993). *Decision mak-ing in action: models and methods.* New York: Ablex Publishing.

Kleinsman, A., & Kaptein, N. (2017). Veiligheid in de GGZ – leren van incidenten en calamiteiten. Diagnosis uitgevers, Leusden, The Netherlands [in Dutch].

Kurtz, C. (2014). *Working with stories in your community or organization: participatory nar-rative inquiry.* New York: Kurtz-Fernhout Publishing.

Kwok, Y. T. A., Mah, A. P., & Pang, K. M. (2020). Our first review: an evaluation of effectiveness of root cause analysis recommendations in Hong Kong public hospitals. *BMC Health Services Research,* 20, 1–9.

Landman, A., Groen, E. L., Van Paassen, M. M., Bronkhorst, A. W., & Mulder, M. (2017). The influence of surprise on upset recovery performance in airline pilots. *The International Journal of Aerospace Psychology,* 27(1–2), 2–14.

Laurence, D. (2005). Safety rules and regulations on mine sites–the problem and a solution. *Journal of Safety Research,* 36(1), 39–50.

Leveson, N. (2011). *Engineering a safer world: systems thinking applied to safety.* Cambridge, MA: The MIT Press.

Leveson, N. (2015). A systems approach to risk management through leading safety indicators. *Reliability Engineering & System Safety,* 136, 17–34.

Leveson, N. G., & Thomas, J. P. (2018). *STPA handbook*. MIT Partnership for Systems Approaches to Safety and Security (PSASS), Cambridge, MA.

Lindhout, P., & Reniers, G. (2017). What about nudges in the process industry? Exploring a new safety management tool. *Journal of Loss Prevention in the Process Industries*, 50, 243–256.

Makary, M. A., & Daniel, M. (2016). Medical error—the third leading cause of death in the US. *BMJ*, 353 (i2139), 1-5. doi: 10.1136/bmj.i2139

Manaadiar, H. (2018). *Lessons learnt from Maersk Honam fire*. Shipping and Freight resource, September 27th, 2018. https://www.shippingandfreightresource.com/lessons-learnt-from-maersk-honam-fire/, accessed October 18th, 2020.

Martin-Delgado, J., Martínez-García, A., Aranaz-Andres, J. M., Valencia-Martín, J. L., & Mira, J. J. (2020). *How much of Root Cause Analysis translates to improve patient safety. A systematic review. Medical Principles and Practice* (in press). doi: 10.1159/000508677

McLeod, R., Berman, J., Dickinson, C., Forsyth, D., & Worthy, T. (2020). *Learning from adverse events*. Chartered Institute of Ergonomics and Human Factors, Birmingham, UK. https://www.ergonomics.org.uk/common/Uploaded%20files/Publications/CIEHF-Learning-from-Adverse-Events.pdf, accessed June 26th, 2020.

Morieux, Y., (2018). Bringing managers back to work. *BCG White Paper*, Boston Consulting Group, October 2018.

Morieux, Y., & Tollman, P. (2014). *Six simple rules: how to manage complexity without getting complicated*. Cambridge, MA: Harvard Business Review Press.

NAM (2018), Onderzoeksrapport Lozing aardgascondensaat op riool en afwateringskanaal te Farmsum (NAM locatie Tankenpark Delfzijl) Oktober 2018, document number EP201901200150; Assen, the Netherlands [in Dutch].

O'donovan, R., & Mcauliffe, E. (2020). A systematic review of factors that enable psychological safety in healthcare teams. *International Journal for Quality in Health Care*, 32(4), 240–250.

Pagell, M., Parkinson, M., Veltri, A., Gray, J., Wiengarten, F., Louis, M., & Fynes, B. (2020). The tension between worker safety and organization survival. *Management Science*, 1–94. doi: 10.1287/mnsc.2020.3589

Parker, A., Ummels, F., Wellman, J., Whitley, D., Groeneweg, J., & Drupsteen-Sint, L. (2018). How to take learning from incidents to the next level. In *SPE International Conference and Exhibition on Health, Safety, Security, Environment, and Social Responsibility*, 1-11. Society of Petroleum Engineers, Richardson, TX. doi:10.2118/190646-MS

Patterson, M. D., & Wears, R. L. (2015). Resilience and precarious success. *Reliability Engineering & System Safety*, 141, 45–53.

Piric, S., Roelen, A., Karanikas, N., Kaspers, S., van Aalst, R., & de Boer, R.J. (2018). How much do organizations plan for a positive safety culture? Introducing the Aviation Academy Safety Culture Prerequisites (AVAC-SCP) tool. *AUP Advances*, 1(1), 118–129.

Plioutsias, A., Stamoulis, K., Papanikou, M., & de Boer, R. J. (2020). Safety differently: a case study in an Aviation Maintenance-Repair-Overhaul facility. *MATEC Web of Conference*, 314(2020), 01002.

Pupulidy, I., & Vesel, C. (2017). The Learning Review: adding to the accident investigation toolbox. In *Proceedings of the 53rd ESReDA Seminar*, ESReDA, Kaunas, Lithuania (pp. 255–263).

Rankin, A., Woltjer, R., & Field, J. (2016). Sensemaking following surprise in the cockpit—a re-framing problem. *Cognition, Technology & Work*, 18(4), 623–642.

Rasmussen, J. (1997). Risk management in a dynamic society: a modelling problem. *Safety Science*, 27(2–3), 183–213.

Reason, J. (1998). Achieving a safe culture: theory and practice. *Work & Stress*, 12(3), 293–306.

Rogers Commission. (1986). Report of the Presidential Commission on the Space Shuttle Challenger Accident. Washington D.C.

Ruitenburg, K. (2017). *Handreiking voorkomen van bijplaatsingen (Guide to prevent misplaced trash)*. Novi Mores, Utrecht, the Netherlands.

Sagberg, F. (2000). *Automatic enforcement technologies and systems*. The Escape Project, Technical Research Centre of Finland (VTT), Espoo, Finland.

Saloniemi, A., & Oksanen, H. (1998). Accidents and fatal accidents: some paradoxes. *Safety Science*, 29, 59–66.

Sasangohar, F., Peres, S. C., Williams, J. P., Smith, A., & Mannan, M. S. (2018). Investigating written procedures in process safety: qualitative data analysis of interviews from high risk facilities. *Process Safety and Environmental Protection*, 113, 30–39.

Semrau, K. E., Hirschhorn, L. R., Marx Delaney, M., Singh, V. P., Saurastri, R., Sharma, N., ... Kodkany, B. S. (2017). Outcomes of a coaching-based WHO safe childbirth checklist program in India. *New England Journal of Medicine*, 377(24), 2313–2324

Service, O., Hallsworth, M., Halpern, D., Algate, F., Gallagher, R., Nguyen, S., ... Sanders, M. (2014). *EAST Four simple ways to apply behavioural insights*. London, UK: Behavioural Insights Team.

Sharygina-Rusthoven, M. O. (2019). The role of context in proactive and voice behavior. Doctoral thesis, Vrije Universiteit Amsterdam, Netherlands, pp. 44, 61, 68, 93, 109.

Sherratt, F., & Dainty, A. R. (2017). UK construction safety: a zero paradox? *Policy and Practice in Health and Safety*, 15(2), 108–116.

Shojania, K. G., & Dixon-Woods, M. (2013). 'Bad apples': time to redefine as a type of systems problem? *BMJ Quality & Safety*, 22(7): 528–531. doi:10.1136/bmjqs-2013-002138.

Shorrock, S. (2020). Learning from behavioural change: a conversation with Nick Godbehere. In Hindsight (30), *European Organisation for Safety of Air Navigation (EUROCONTROL)*, Spring 2020 (pp. 70–75). EUROCONTROL, Brussels.

Shorrock, S., Wennerberg, A., & Licu, T. (2018). Competency and moral dilemmas: what would you do? In Hindsight (27), *European Organisation for Safety of Air Navigation (EUROCONTROL)*, August 2018 (pp. 41–44). EUROCONTROL, Brussels.

Simon, H. A. (1955). A behavioral model of rational choice. *The Quarterly Journal of Economics*, 69(1), 99–118.

Simons, D. J., & Chabris, C. F. (1999). Gorillas in our midst: sustained inattentional blindness for dynamic events. *Perception*, 28(9), 1059–1074.

Snowden, D. J., & Boone, M. E. (2007). A leader's framework for decision making. *Harvard Business Review*, 85(11), 68.

Storkersen, K., Antonsen, S., & Kongsvik, T. (2016). One size fits all? Safety management regulation of ship accidents and personal injuries. *Journal of Risk Research*, 20(7), 1–19.

Sunshine, J. E., Meo, N., Kassebaum, N. J., Collison, M. L., Mokdad, A. H., & Naghavi, M. (2019). Association of adverse effects of medical treatment with mortality in the United States: a secondary analysis of the Global Burden of Diseases, Injuries, and Risk Factors study. *JAMA Netw Open*, 2(1), e187041. doi:10.1001/jamanetworkopen.2018.7041

Tesselaar, S., & Rodermond, J. (2017). De Gids: Narratief Evalueren. Eburon, Delft, p. 99 [in Dutch].

Thomas, E. J. (2020). The harms of promoting 'Zero Harm'. *BMJ Quality & Safety*, 29, 4–6.

Tromp, N., & Hekkert, P. (2018). *Designing for society: products and services for a better world*. London and New York: Bloomsbury Publishing

Tversky, A., & Kahneman, D. (1973). Availability: a heuristic for judging frequency and probability. *Cognitive Psychology*, 5(2), 207–232.

UMC Utrecht. (2019). Reflectie raad van bestuur op SIRE-Rapport. May 7th, 2019, available at https://www.umcutrecht.nl/getattachment/7bde2185-b565-451a-a445-c0788271b532/SIRE_reflectie_rvb_190507.pdf.aspx?lang=nl-NL, accessed February 26th, 2020.

Vaughan, D. (1999). The dark side of organizations: mistake, misconduct, and disaster. *Annual Review Sociology*, 25, 271–305.

Vlasblom, J. I., Pennings, H. J., van der Pal, J., & Oprins, E. A. (2020). Competence retention in safety-critical professions: a systematic literature review. *Educational Research Review*, 30, 1-14. doi: 10.1016/j.edurev.2020.100330

Weeda, F. (2018a). Niemand was in mij geïnteresseerd. NRC Handelsblad, April 13th, 2018. https://www.nrc.nl/nieuws/2018/04/13/niemand-was-in-mij-geinteresseerd-a1599431/appview

Weeda, F. (2018b). Zwijgen was voor Adrienne Cullen geen optie. NRC Handelsblad, December 31st, 2018. https://www.nrc.nl/nieuws/2018/12/31/zwijgen-was-voor-adrienne-cullen-geen-optie-a3127503/appview

Weeda, F. (2020). Strijdbare patiënte Cullen kreeg geen informatie over onderzoek medische misser. NRC Handelsblad, January 2nd, 2020. https://www.nrc.nl/nieuws/2020/01/02/strijdbare-patiente-cullen-kreeg-geen-informatie-over-onderzoek-medische-misser-a3985507/appview

Weick, K. E., Sutcliffe, K. M., & Obstfeld, D. (2005). Organizing and the process of sensemaking. *Organization Science*, 16(4), 409–421.

Woods, D., & Hollnagel, E. (2006). *Joint cognitive systems: patterns in cognitive systems engineering*. Boca Raton, FL: Taylor & Francis.

Zehr, H. (2015). *The little book of restorative justice: revised and updated*. New York: Simon and Schuster.

Zohar, D., & Marshall, I. N. (1994). *The quantum society mind, physics and a new social vision*. New York: HarperCollins.

Glossary

Adverse event: An event that could have caused or did result in harm to people, including psychological harm, feelings of hurt, injury or death, damage to property or reputational harm. We try to prevent adverse events in the interest of safety. It includes accidents, incidents and near-misses. Adverse events are a natural consequence of the complexity of the organisation and its relationship with its environment.

Appreciative Inquiry: A change management approach that focuses on what is working well analysing why it is working well and then doing more of it.

Appreciative Investigations: Guided interactions of managers and leaders with field experts to understand how people create successful outcomes. A form of Appreciative Inquiry.

Automation Surprise: When 'crews are surprised by actions taken (or not taken) by the automated system'. See Woods and Hollnagel (2006, p. 119).

'Bad News': Information about gaps between rules and practice, or the fact that a near-miss occurred, that might damage your reputation but should be welcomed as a chance to learn and improve.

Blunt end: The organisation or set of organisations that both support and constrain activities at the sharp end, isolated in time and location from the event. For instance, the planning and management actions related to a task. The opposite is the sharp end: There where people are in direct contact with the safety-critical process, and therefore closest in time and geography to the (potential) adverse event. See Dekker (2014, p. 39).

Bounded rationality: The suggestion that humans do not fully optimise when making decisions, because of limits in cognitive powers and available time. Leads to satisficing behaviour. See Simon (1955).

Buggy mental model: When automation seems untrustworthy from the user's perspective because the automation logic does not match the operator's sense of what should be happening. See Dekker (2014b, p. 99).

Clutter: The accumulation and persistence of safety procedures and activities that are performed in the name of safety, but do not contribute to the safety of operational work. See Dekker (2017, p. 193).

Complex system: A system that shows *emergent* behaviour that we cannot predict but that is understandable in retrospect. It is distinct from ordered systems in that those show predictable behaviour (for everyone in the case of simple ordered systems and for experts for complicated systems). Complex systems are also distinct from chaotic systems where the relationships between cause and effect are impossible to determine because they shift constantly, and no manageable patterns exist. A complex system has the following characteristics: it involves large numbers of interacting elements; the interactions are nonlinear, and minor changes can produce disproportionately major consequences; the system is dynamic, the whole is greater than the sum of its parts, and solutions can't be imposed – rather, they arise from the circumstances; the system has a history, and the past is integrated with the present, the elements evolve with one another and with the environment, and evolution is irreversible; though a complex system may, in retrospect, appear to be ordered and predictable, hindsight does not lead to foresight; we cannot forecast or predict what will happen. See: Snowden and Boone (2007).

Continuous improvement opportunity: What used to be called 'an audit', but now focused on maximising learning rather than minimising findings. See Brown (2019).

Contribution of protective structures: A factor in drift that makes the (mal) functioning of the system opaque. It includes assurance structures (regulatory arrangements, quality review boards, safety departments), assurance information (reporting systems, meeting notes, databases) and staffing choices (previous experience, affiliations, ambitions, renumeration and social bonds). See Dekker (2011).

Counterfactual descriptions: Things that did not occur but could have occurred, for example 'The crew did not monitor the aircraft's position using a ground movement chart'. Counterfactual descriptions do not help us understand why the adverse event occurred (in this case, a take-off from a taxiway instead of the runway). Rather, we need to understand how this aircraft crew usually navigates around their home airport and whether their way of doing it is common practice. Counterfactuals are sometimes a result of too much faith in Work-as-Imagined and an indication that the gap with Work-as-Done has not been sufficiently investigated.

Decluttering: Reducing Clutter (see there).

Decrementalism: The tendency to judge risks as more acceptable if the hazard is introduced in small steps over time rather than one big change – a factor in drift. See Dekker (2011).

Design Thinking: 'a human-centred process of finding creative and innovative solutions to problems' that can be used to support the success of micro-experiments. 'By approaching the process using design methods, organisations and teams in any field can better understand their users, redefine challenges, and quickly test and iterate on possible solutions. [...] The design thinking process has five stages that can be approached in both a linear and non-linear fashion. The five stages of Design Thinking are: Empathize, Define, Ideate, Prototype, and Test'. See https://www.sessionlab.com/blog/design-thinking-online-tools/#design-thinking-faq, accessed June 16th, 2020.

Drift (into failure): Eroding safety margins and an increase in risky behaviour that is inevitable in any organisation in the absence of serious incidents. Drift is driven by five factors: scarcity and competition, decrementalism, sensitivity to initial conditions, unruly technology and failing protective structures. See Dekker (2011).

Edges of the operating envelope: A term to describe that operations are only just within prescribed limits, regarding the weather, available resources, process conditions, etc. This is where safety margins are considered to be the smallest. See Dekker (2011, p. 178).

Efficiency–Thoroughness Trade-Off (ETTO): The choice that people make in leisure or at work between being thorough or efficient, since it is rarely possible to be both at the same time. The trade-off is driven by demands for productivity versus safety. See Hollnagel (2009).

Emergent behaviour: Traditionally, this is behaviour of a system that is not apparent from its components in isolation, but which result from the interactions, dependencies or relationships they form when placed together in a system. In non-complex cases, this behaviour is intended or expected, for example the movement of a bicycle when pedalled by a human. In complex systems, emergent behaviour can often not be entirely predicted (although it is often explainable in hindsight), because small causes can lead to exorbitant effects (called the 'cause–effect asymmetry' or the 'butterfly effect'); complete knowledge of a complex socio-technical system is impossible because of the open nature of the system and its resultant sensitivity to triggers from the environment; and complex systems demonstrate history or path dependence: it matters how a certain situation was arrived at. See Kaspers, Karanikas, Roelen, Piric and Boer (2019); Zohar and Marshall (1994); Dekker, Cilliers and Hofmeyr (2011).

Finding (in an investigation): The discovery and establishment of the facts of an issue. The word 'discovery' suggests that the investigator needs to actively pursue relevant findings. 'Establishment' suggests that credibility for the finding needs to be constructed before it can be interpreted as a fact. The phrase 'of an issue' suggests that there will be findings that are not relevant for understanding the event under consideration and should therefore be discarded. Findings can be 'facts that have been verified, presumptions that have been proven to be true or false (based on the available facts or analysis), or judgements beyond reasonable doubt when dealing with human and organisational factors'. See European Safety Reliability and Data Association (ESReDA) (2009, p. 33).

First victims: Victims suffering the direct consequences of the event without having played an active role in its cause. Examples include patients in healthcare, passengers in transport and the inhabitants living close to a process plant. Often includes the families of these as well. See Dekker (2016).

Freedom in a frame: The condition that operators feel trusted to make the right choices when they are executing their tasks, and the need to deviate from the procedures is signalled to peers or supervisors. Front-line employees feel responsible for adherence to the procedures and deviate from them at their own discretion but knowingly and justly. They take responsibility for these exemptions and are willing and able to explain their considerations. There is effective peer pressure to abide by the rules unless the circumstances dictate otherwise. It stands to reason that this is only possible if they accept Work-as-Imagined as a useful basis from which to assess what needs to be done in practice.

HILF (or HILP) events: High-impact low-frequency (high-impact low-probability) events that are potentially catastrophic but occur very rarely, and are therefore not usually represented in the experience of employees or managers, such as explosions or aircraft crashes.

Hindsight bias: See retrospective biases.

'Human Error': Retrospective designation for an action by an actor that leads to an adverse outcome. Generally, similar actions routinely lead to successful outcomes, and so the identification of a 'human error' requires us to delve deeper and determine the causes for this action and/or its outcome. The quotation marks represent the fact that an attribution to 'human error' can only occur after the outcome of the event is known. See Dekker (2014b).

Human Factors (or Ergonomics): 'is the scientific discipline concerned with the understanding of interactions among humans and other elements of a system, and the profession that applies theory, principles, data, and other methods to design in order to optimize human well-being and overall system performance'. See https://iea.cc/what-is-ergonomics/.
'Ergonomics and human factors use knowledge of human abilities and limitations to design systems, organisations, jobs, machines, tools, and consumer products for safe, efficient, and comfortable human use'. See https://www.surfnetkids.com/tech/1355/computer-ergonomics-for-elementary-school-students/

'Just Culture' (traditional): Approach currently common in industry that aims to differentiate between acceptable and unacceptable performance and uses this to determine appropriate consequences. It is often manifest in flow charts that support someone in determining how culpable a 'perpetrator' is. A traditional 'Just Culture' is more focused on who is responsible than learning about the factors that led to the event, potentially leaving these unaddressed. A traditional 'Just Culture' tends to consider the existing procedures and rules as legitimate – possibly involving the rule makers or rule owners in the evaluation of the 'offenders', and even though the rules may not have been suitable in the context of the event. And whereas a traditional

'Just Culture' in organisations seems to mirror justice in society at large, the crucial elements of an independent judge and the right to appeal do not exist. Not coincidentally, managers believe their organisation to be more just than those lower in the hierarchy, leading to a lack of reports, honesty, openness, learning and prevention – hence the quotation marks. See Dekker (2016).

Learning review: An accident investigation technique that explores the human contribution to accidents, safety and normal work in complex environments. More than traditional investigations, it is designed to facilitate the understanding of the factors and conditions that influence human actions and decisions by encouraging individual and group sensemaking at all levels of the organisation. 'The Learning Review introduces the need to create a narrative inclusive of multiple perspectives from which a network of influences map can be created. This map depicts the factors that influence behaviours and can aid the organisational leadership to effect meaningful changes to the conditions while simultaneously helping field personnel to understand and manage system pressures'. See Pupulidy and Vesel (2017). Four good (and free at the time of writing) resources to learn from events are as follows: European Safety Reliability and Data Association (ESReDA 2009, 2015); Allspaw, Evans, Schauenberg (2016); and McLeod, Berman, Dickinson, Forsyth and Worthy (2020).

Micro-experiments: Small initiatives to try out improvements around a particular topic in a safe-to-fail manner, meaning that if the intervention is unsuccessful, it does not lead to disproportional damage. Means to amplify (expand) in case it is successful or dampen (discontinue) it need to be considered before commencing. Generally based on ideas that are generated bottom-up by front-line employees, and coaching and facilitation make for more, and more successful, micro-experiments. The design-thinking process can be used to support the success of micro-experiments.

Narratives: Short stories that have been experienced by the storyteller himself/herself and that can be collected through an app on a smart phone or tablet. The system probes for examples of where work was difficult and through a number of follow-on questions is able to categorise the example or add meaning. See Kurtz (2014); de Boer (2016). The app that we used was made available by Storyconnect, https://storyconnect.nl

New View (of 'human error'): Behaviour which we call 'human error' is a consequence of trouble deeper inside organisations. The conditions surrounding an adverse event need to be investigated, rather than assuming that people have come to work to do a bad job. Incorporated in the Safety Differently approach and opposite of 'Old View'. See Dekker (2014b, p. xv).

No-blame culture: '... recognises that operating problems are often complex and usually are not the fault of only one person. So where an operating problem has occurred, the business is far better off finding out what happened. The first step to achieving this is to stop looking for the culprit. Though there may be the odd person who comes to work intending to fail, the vast majority of us intend to succeed and as managers, our job is to help them, not discipline them'. (Hand 2016, p. 2). However, such a no-blame culture disregards the obligation that practitioners have to restore relations and distil learning from an incident, and is therefore an inferior alternative to Restorative Practice.

Normalisation of deviance: Normalisation of deviance means that people within the organisation become so accustomed to deviant behaviour that they don't consider it as deviant, despite the fact that they exceed their own rules for safety. See Vaughan (1999).

Nudging: Manipulation of choices that are made unconsciously by an operator, for instance, to ensure more compliance with good practice, unwritten cultural rules or written instructions. Lindhout and Reniers (2017) have identified nine types of nudges for safety management.

Old View (of 'human error'): Traditional, common and ineffective view of human performance that suggests that 'human error' needs to be eradicated by correcting and controlling behaviour. Opposite of 'New View'. See Dekker (2014b, p. xv).

Outcome bias: See retrospective biases.

Psychological Safety: Circumstances in which people are not hindered by interpersonal fear and are willing to be candid (including sharing concerns and mistakes without the fear of retribution), so that team and organisational performance is maximised. See Edmondson (2018).

Resilience: Resilience is 'the intrinsic ability of a system to adjust its functioning prior to, during, or following changes and disturbances, so that it can sustain required operations under both expected and unexpected conditions' (Hollnagel 2013, p. xxxvi). This relies on the capacity of individuals, teams and organisations to adapt to a threat or a changing work environment, and draws attention to the ingenuity and adaptability that professionals display to maintain ordinary and apparently "standard" operations under challenging and variable conditions. Four abilities support resilience: how a system responds, how it monitors, how it learns and how it anticipates. These abilities are developed through a continuous discussion on risk and ensuring that operators develop expertise to operate at the edges of the envelope. See Amalberti & Vincent 2020.

Restorative Practice (also called 'Restorative Just Culture'): A process aimed at restoring relations after an adverse event. It asks who was hurt, what are their needs and whose obligation is it to meet those needs. It differs from 'Just Culture' in that the latter is aimed at retribution and works on the basis of an algorithm that requires someone to determine how culpable a 'perpetrator' is. We have found Restorative Practice to be very effective within organisations to reduce blame, enable transparency, improve motivation and therefore also create economic benefits. Sometimes also called 'Restorative Just Culture', but this is confusing as it is quite different to a traditional 'Just Culture'. See Dekker and Breakey (2016); Dekker (2016). Note: the third edition is quite different to earlier versions and better explains Restorative Practice. Zehr (2015).

Retrospective biases: Hindsight bias and outcome bias. These are closely related – hindsight bias overestimates one's ability have predicted an outcome; outcome bias unfairly judges a decision on information only available after the decision's outcome is known.

Root-cause analysis: Synonym for Learning Review that is particularly common in healthcare globally. Although it is aimed at 'uncovering the systems-level *causes* and contributing *factors* behind an incident or near-miss' [emphasis added], the use of 'root' and the singular 'cause' in the title may obscure that a multitude of factors precede an adverse event. See Card, Ward, and Clarkson (2012).

Safety: Traditionally regarded as described under Safety I, but more recently as described under Safety II.

Safety I: Traditional view in which safety is considered as the absence of accidents and incidents (backward looking) or as an acceptable level of risk to inflict damage (forward looking). 'In this perspective, safety is defined as a state whereas few things as possible go wrong. A Safety-I approach presumes that things go wrong because of identifiable failures or malfunctions of specific components: technology, procedures, the human workers and the organisations in which they are embedded. [...] Humans – acting alone or collectively – are therefore viewed predominantly as a liability or hazard, principally because they are the most variable of these components'. See Hollnagel, Wears, and Braithwaite (2015).

Safety II: 'The system's ability to function as required under varying conditions, so that the number of intended and acceptable outcomes is as high as possible (as is usually the case, on all those days that we do not experience an adverse event). The basis for safety and safety management must therefore be an understanding of why things go right, which means an understanding of everyday activities. Usually depicted as a Gaussian (bell-shaped) probability distribution, suggesting that not just the unwanted outcomes but all events are relevant to safety'. See Hollnagel, Wears, and Braithwaite (2015).

Safety Culture: A pattern of shared basic assumptions that the group learned as it solved its problems of external adaptation and internal integration, that has worked well enough to be considered valid and, therefore, to be taught to new members as the correct way to perceive, think and feel in relation to safety (Adapted from Guldenmund 2000, pp. 11, 37). Somewhat ambiguous term that invites much theoretical debate but has limited practical value. Like any culture, safety culture is inherently stable (requiring multiple years for noticeable changes), and it is ineffective to tweak just single elements of a culture. Focusing on culture means that many years will pass before a change will be perceptible, and in the meantime, the results will be confounded by other internal and external influences. Ritually, a safety culture survey might be rolled out, but these have very limited predictive value regarding incidents. Depending on the exact definition, an advantageous safety culture is just as much a consequence of operational improvements, as it can be the cause. For instance, to suggest that 'safety culture' is a factor in improving psychological safety is a tautology. Indicators of both 'safety culture' and psychological safety are that staff (including novices) can speak up and discuss concerns openly. See for an example of such a tautology: O'donovan and Mcauliffe (2020). Although some non-scientific sources suggest 'safety culture has been shown to be a key predictor of safety performance' (https://www.pslhub.org/learn/culture/good-practice/nes-safety-culture-discussion-cards-r2506/), science comes no further than a proxy (i.e. correlation). Some suggest that a discussion on Safety Culture can enhance safety (e.g. https://www.skybrary.aero/index.php/File:Safety_culture_discussion_cards_-_Edition_2_-_20191218_-_hi_res_(without_print_marks)_Page_02.png), but this seems just as roundabout as talking about profitability to improve company profits. It can help create understanding but doesn't carry much weight in actual performance improvement. See Dekker (2019, pp. 363–387); Guldenmund (2000, pp. 11, 37); Henriqson, Schuler, van Winsen and Dekker (2014, p. 474); O'donovan and Mcauliffe (2020).

Safety Differently: A relatively recent safety paradigm that has proven effective in many industries but is controversial because it rejects the term 'human error', challenges the established culture of absolute compliance, suggests that 'zero-harm' policies are counterproductive and requires the acceptance of accidents as part of working life. For an introduction to Safety Differently, see Dekker (2014a, 2019) and Doing Safety Differently (the movie), available at https://www.youtube.com/user/sidneydekker, accessed April 16th, 2020.

Safety Management System: A systematic approach to managing safety, including the necessary organisational structures, accountabilities, policies and procedures. In domains like aviation, such a system is required for many companies in an industry-wide effort to improve safety even further. A safety management system typically includes components like policy and objectives, risk management, assurance and promotion. Recent developments are suggesting that the safety management system will migrate to a management system, removing the artificial segregation between safety and other management realms. Although a (safety) management system can lay a foundation for useful policies and effective processes, any safety effort is wasted if it is aimed at substituting bureaucratic compliance for chronic unease, and metrics for understanding. A vital component of contemporary safety management systems is risk identification. But there is little that you can do to manage and mitigate a risk that you don't know about or have judged incorrectly. See ICAO (2013a, 2013b) or https://www.skybrary.aero/index.php/Safety_Management_System, accessed August 23rd, 2018..

Safety rituals: (often obligatory) actions like holding bannisters, parking the car backwards and refraining from phoning hands-free while driving that do not significantly contribute to the reduction of adverse events but are thought to contribute to an effective safety culture. See Hudson (2007).

Satisficing behaviour: A decision-making strategy to choose an alternative that satisfies minimal criteria rather than to calculate the optimal option. This approach minimises the cognitive resources that are required and takes into account lack of perfect information; and is thought to be the default behaviour for humans. See also bounded rationality. See Simon (1955).

Scarcity and competition: A factor in drift caused by economic pressures leading to a reduction in staffing levels and quality of resources. See Dekker (2011).

Second victims: Practitioners involved in an adverse event who feel personally responsible and suffer as a result of this. See Dekker (2016).

Self-verification: Conversations between managers and operators intended to identify issues and gaps between Work-as-Imagined and Work-as-Done, hopefully in an open and engaging manner to be most useful. These conversations can be ad hoc or scheduled.

Sensemaking: A process for turning circumstances into a situation that is comprehended explicitly in words and that serves as a springboard for action. Sensemaking is similar to analysis in that it is focused on taking decisions that are pertinent to a specific situation and is grounded in concrete reality. But it differs from ordinary analysis in that sensemaking happens on the fly without predefined procedures, and it is often executed collaboratively with different interpretations of the situation appearing in quick succession. Sensemaking focuses on plausibility rather than accuracy and facilitates retrospection. It is these differences that make organisational sensemaking more powerful in a complex environment than analysis. See Kurtz (2014); Rankin, Woltjer and Field (2016); Weick, Sutcliffe and Obstfeld (2005).

Sensitivity to initial conditions: A factor in drift where the safety of the original design and implementation is compromised over time by changes in circumstances and modifications to the system. See Dekker (2011).

Sharp end: There where people are in direct contact with the safety-critical process, and therefore closest in time and geography to the (potential) adverse event. The opposite is the blunt end: the organisation or set of organisations that both support and constrain activities at the sharp end, isolated in time and location from the event. For instance, the planning and management actions related to a task. See Dekker (2014b, p. 39).

Synesis: A new term introduced by Hollnagel (2019) to represent 'the condition where practices are brought together to produce the intended outcomes in a way that satisfies more than one priority and possible [sic] reconciles multiple priorities and also combines or aligns multiple perspectives'. Synesis is needed as a replacement for 'Safety II' because 'the meaning of the word "safety" in Safety II has little to do with the traditional interpretation of safety'. See Hollnagel (2019).

Tipp-ex accidents: In safety science not a mishap that leaves a white, difficult to remove stain but injuries that should be reported or counted, but that are swept under the carpet in an effort to retain an unblemished safety record. See https://tippexongeval.be/

Unruly technology: Technology (often of the information-processing kind) that has passed quality assurance yet displays unexpected behaviour in practice. A factor in drift. See Dekker (2011).

Violation: Normative wording that describes a deviation from known rules, policies or procedures, even if they are not fit for purpose. Use of neutral, non-normative synonyms as 'exception' is recommended to facilitate hearing about gaps between *paper* and *practice*.

'Voice' (noun, as in voicing concern): 'a form of organisational behaviour that involves "constructive change-oriented communication intended to improve the situation," even when others disagree. Voice serves as a "seed corn for continuous improvement" and allows employees to channel their dissatisfaction with the status quo. Voice behaviour is focused on correcting mistakes, improving processes and formulating solutions to organisational problems'. Voice is generally threatening for the recipient, even if the content itself is not, because it challenges authority. The effect of 'promotive' voice (ideas about improvements, solutions and possibilities) and 'prohibitive' voice (expressions of concern, and harmful factors) on the leader's reaction differs. See Sharygina-Rusthoven (2019).

Work-as-Done (abbreviated WaD, or in this text as *practice*): How work is actually executed, what people have to do to get the job done in the actual situation. People adjust what they do to match the situation such as lack of resources (time, manpower, materials, information, etc.), conflicting goals, missing expertise or motivation.

Work-as-Done is 'in the moment', that is to say it is being executed at a specific place and time and isolated from reflection on it before or after the fact. Work is done many times over, so new instances of Work-as-Done are created all the time.

Work-as-Imagined (abbreviated as WaI, or in this text as *paper***):** How a task is prescribed in rules and procedures. Needed to help people remember the multiple steps of a task, to help educate and train people for their jobs, to ensure that people can cooperate effectively, for design and planning purposes within an organisation, and as a means to identify variances in behaviour that lead to unacceptable risks.

Index

Note: Page numbers in **bold** denote Glossary.

Printed in the United States
By Bookmasters